CONSERVATION AND SUSTAINABLE DEVELOPMENT

The links between policy and practice in natural resource management are often depicted as a cyclical and rational process. In reality, policymaking and implementation are often irrational, unpredictable and highly political. Many science and knowledge-based institutions undertake rigorous research with the aim of influencing policy, but often their influence is much less than intended. Understanding who influences policy at different levels, and how, is crucial to ensure that science is deployed most effectively so as to have an influence on conservation and natural resource management.

Conservation and Sustainable Development presents a variety of innovative ways that have been used to influence policy processes, from community pressure groups through elected and unelected leaders, to scientific discourse at the levels of directors of economic planning and conservation. This book analyzes experiences from a variety of conservation interventions by the International Union for Conservation of Nature (IUCN) and other agencies, primarily in Eastern Africa, and challenges the notion of policymaking as a cyclical process. It elaborates on this theme and presents an array of examples of how communities have influenced government, through direct lobbying, influence of parliamentarians, wielding of science and research, and inter-community dialogue, networking and solidarity. The authors present a framework for understanding and strategizing such work so that other institutions can identify where they can best add value.

Jonathan Davies is Coordinator of the Global Drylands Initiative of the International Union for Conservation of Nature, based in Nairobi, Kenya.

CONSERVATION AND SUSTAINABLE DEVELOPMENT

Linking Practice and Policy in Eastern Africa

Edited by Jonathan Davies

International Development Research Centre
Ottawa • Cairo • Dakar • Montevideo • Nairobi • New Delhi • Singapore

Routledge
Taylor & Francis Group
LONDON AND NEW YORK

IUCN

earthscan
from Routledge

First published in paperback 2024

First published 2012
by Routledge
4 Park Square, Milton Park, Abingdon, Oxon OX14 4RN

and by Routledge
605 Third Avenue, New York, NY 10158

Earthscan is an imprint of the Taylor & Francis Group, an informa business

Co-published with the
International Development Research Centre
(IDRC publishes an ebook edition of this book, ISBN 978-1-55250-533-5)

British Library Cataloguing in Publication Data
A catalogue record for this book is available from the British Library

Library of Congress Cataloging in Publication Data
Conservation and sustainable development : linking practice to policy in Eastern Africa / edited by Jonathan Davies.
p. cm.
Includes bibliographical references and index.
1. Conservation of natural resources—Government policy—Africa, Eastern. 2. Conservation of natural resources—Africa, Eastern—Citizen participation. 3. Sustainable development—Africa, Eastern. I. Davies, Jonathan, Dr. II. International Development Research Centre (Canada)
S934.A354C66 2012
333.72096875—dc23 2011049150

ISBN: 978-1-84971-404-4 (hbk)
ISBN: 978-1-03-292666-7 (pbk)
ISBN: 978-0-203-13148-0 (ebk)

DOI: 10.4324/9780203131480

Typeset in Bembo
by Florence Production Ltd, Stoodleigh, Devon

CONTENTS

FIGURES AND BOXES

Figures

Boxes

FOREWORD

As 2015 approaches – the date set to meet the Millennium Development Goals – we are acutely aware that the joint challenges of reducing poverty and ensuring environmental sustainability are greater than ever. In 2011 famine again rears its ugly head in the Horn of Africa, reflecting the extent to which poverty remains entrenched and the high degree of food insecurity in the region. Climate change is becoming more evident, further burdening populations that are already struggling to adapt and develop. Underlying all of this is an ongoing degradation of natural resources that threatens the development processes and entrenches poverty in much of the region.

Our development depends on nature, and if development in our region is to proceed it must be based on sustainable management of the ecosystems that support life. Although this simple fact is well accepted, achieving sustainable development in practical terms is very challenging. Policy-makers need to make the best decisions they can based on the facts that are available to them, but often the facts at their disposal are inadequate, the research is inconclusive, or the policy options are poorly evaluated. Alternatively the people who hold the answers for policy-makers – farmers, pastoralists and other natural resource users – are not connected to them or lack the skills and opportunities to make their case and share their experiences.

This book shows that there are genuine opportunities to achieve positive development and environment outcomes together. Environmental concerns do not have to displace development priorities, but in order to achieve these win-wins it is necessary to rethink the way environment and development are planned. Translating locally proven good practices into nationally supported policies and investments requires strong partnerships at different levels to improve communication and evaluation of experiences, and to identify policies that really work.

To influence both policy and practice effectively it is necessary to have strong partnerships with a wide array of actors. Communities need to be involved to enrich the policy process with their experiences and knowledge from the ground. Their participation in policy and planning is vital to ensure their ownership and participation in policy implementation. Government at different levels must be involved to ensure legitimacy and public support, and to ensure cohesion between different sectors that impact on natural resources. Leaders, traditional and elected, need to be engaged in policy and planning to ensure wider respect for the process and wider understanding of the outcomes of policy.

What this book contributes is a fresh look at the diverse array of actors who engage in policy and planning at different levels. The book helps to understand the role these different actors can plan, the support they may need, and how they can collaborate to achieve mutually desirable outcomes. Understanding and harnessing the different roles will greatly strengthen policy formulation and planning processes, and will contribute to more effective and sustainable development in the long-term.

The chapters in this book present the experiences of a variety of experts, including government staff, communities and researchers, in trying to feed relevant experiences into government decision-making. The chapters explore the nature of these decision-making processes, and the great diversity of opportunities that can be taken to influence policy and planning. The book explores the roles of each group of actors in shaping policy and planning: communities, technocrats, elected and non-elected leaders, scientists. The book also explores the great breadth of communication tools that can be used to transmit messages to different audiences. The chapters present an array of examples and case studies to illustrate many different approaches.

The book concludes with some practical advice on how to use these different approaches: how to influence different targets, how to present different messages, and how to understand the strengths of your own institution to engage in influencing policy and planning. As such the book has something for everyone. A key message is that, for all these different actors to influence policy and planning effectively, they need to be much more aware of the complexities of decision-making processes, and they need to work in partnership to share their strengths in order to have the greatest impact. Ensuring the most strategic alliances is an important first step.

Communities and their representatives therefore can learn a great deal from this book about how to present their ideas and experience effectively. Community Groups and NGOs similarly will benefit from the approaches and planning tools that the book presents. Policy-makers themselves – the target of much advocacy – may find this book informative for improving the quality of their consultations and ensuring greater understanding and ownership of decision-making processes.

Eng. Mahboub Maalim
Executive Secretary, the Intergovernmental Authority on Development

NOTES ON CONTRIBUTORS

Jonathan Davies is the Coordinator of the IUCN Global Drylands Initiative, within the Ecosystem Management Programme. He has lived in Eastern Africa since 1999, working initially in the field of humanitarian emergencies and later in sustainable development and conservation. He has engaged in policy advocacy with communities, local and national government and international organizations for the past 16 years, through work in Asia and Africa. He led the development of this book between 2008–2011 when he was part of the Technical Coordination Group of IUCN in Eastern and Southern Africa.

Frits Hesselink, born in the Netherlands in 1945, is a former Chair (1994–2000) of the IUCN Commission on Education and Communication (CEC). Currently he is an Advisor to the CEC Chair. He has a background in international law and worked since 1975 in the field of communicating conservation and sustainable development. Since 1997 he has worked as international consultant for environmental conventions, international organizations, governments, NGOs and the private sector in various parts of the world (www.hect.nl). To explore what makes him tick in his day-to-day work on 'deep listening to facilitate positive change', read his blog: the Art of Positive Change (http://cepatoolkit.blogspot.com).

Victor Kamagenge worked for IUCN, the International Union for Conservation of Nature as Programme Officer, based in Dar Es Salaam, Tanzania. He has a background in Water Resources and Environmental Engineering and has worked on issues related to community empowerment in conserving the natural resources and livelihood improvement for over ten years with International Organizations and NGO as staff. This includes experience in working in the conservation of the natural resources within the Nile Basin as well as the Marine resources in Tanga coastal area in Tanzania.

Sophie Kutegeka works for IUCN as a Programme Officer based in Kampala, Uganda. Sophie has experience in sustainable natural resource management with a core focus on climate change, forestry and extractives. She gained a lot of experience in natural resource governance while serving as a Programme Officer at the Advocates Coalition for Development and Environment (ACODE), a policy research and advocacy think tank based in Uganda. In 2009, she served as a capacity advancement fellow with Revenue Watch Institute, New York undertaking research and advocacy to promote transparency and accountability in the emerging oil and gas sector, and its implications on the environment and the economy in Uganda. She serves as an adviser on the East African Board for Global Green Grants Fund where she supports community groups to implement conservation projects, which help to improve their livelihoods.

Pablo Manzano Baena works for the International Union for Conservation of Nature (IUCN), as the Global Coordinator of the World Initiative for Sustainable Pastoralism (WISP). A Spaniard by nationality, Pablo is a Rangeland Ecologist who has worked with pastoralism-related issues during the last ten years, both in the field of scientific research and international development. His international experience includes Spain, Argentina, Germany and Bosnia and Herzegovina. In his current position, based in Nairobi, Kenya, he works at influencing policies from the local to the global levels through mainstreaming the understanding of pastoralism with evidence-based arguments.

Barbara Nakangu Bugembe has worked for 11 years implementing development programs in Agriculture and Environmental Conservation and rural development. She currently heads the IUCN office in Uganda. Prior to joining IUCN, she worked in the Uganda National Council for Science and Technology as a resident consultant and as a project officer with a local NGO known as FIT Uganda. Within IUCN Barbara has worked on transboundary projects in Mount Elgon, wetlands projects, and climate change projects focusing on both Adaptation and Mitigation all of which aimed to influence policy based on field-based evidence. Barbara holds a MSc in Agricultural economics and a BSc in Agricultural sciences.

Claire J. Ogali works for IUCN as a Programme Assistant for the Global Drylands Initiative, based in Nairobi, Kenya. She has a background in Environment and Natural Resource Management. With four years of work experience, Claire looks forward to an enriching career in the conservation world and policy influence at the local, national, regional and global level. Prior to joining IUCN, she worked with the Aga Khan Foundation East Africa.

Guyo M. Roba is Programme Officer for the IUCN Drylands Programme based in Nairobi, Kenya. He has a background in environment, energy and public policy studies and has worked on issues related to Energy and Natural Resource management for over seven years. This includes experience in Kenya and other Eastern Africa countries, with half of his career spent in providing research-based public policy advice and capacity building to the government of Kenya, local communities and to the private sector in order to contribute to achievement of development and conservation goals. He has worked as policy analyst in Kenya and has influenced policy and practice at different levels on infrastructure, environment and energy sector. Accordingly, he has trained, written and published in the areas of infrastructure, resource and environmental management and public policy.

Ben Wandago is a Natural Resource Manager with over 22 years' experience. He has a BSc in Forestry Science from Moi University – Kenya and MSc in Environmental Science from Wageningen Agricultural University, The Netherlands. He was previously employed by the Kenya Forest Department (now known as the Kenya Forestry Service) where he served in various capacities and worked extensively on forest policy formulation and implementation at different levels. He also worked as a Senior Program Officer with IUCN Eastern and Southern Africa Regional Office for four years with the overall responsibility of Technical and Managerial support to the Drylands Programme. Currently he works for the African Wildlife Foundation as the Samburu Heartland Coordinator.

Juliane Zeidler is co-founder of Integrated Environmental Consultants Namibia (IECN) and Namibia-based Nature Trust. She has been working in the fields of environment and development for more than 15 years and is well-known, in Namibia and internationally, for her expertise in biodiversity research, climate change adaptation (especially at the community level), natural resources management, community development, environmental politics and sustainable development. Juliane is the Regional Vice Chair of the IUCN Commission for Education and Communication (CEC) and has a specific interest in communication, outreach and capacity development for conservation and sustainable development, especially in Africa. She finds that much knowledge and research information are being generated but are mostly only haphazardly being used for strategic communications and engagement in policy processes – which often are quite informal in nature. Juliane advocates for a stronger focus on these important aspects of policy advocacy work, as they can add real value to our international conservation and development efforts.

ACKNOWLEDGEMENTS

IUCN, The International Union for Conservation of Nature, would like to acknowledge the generous support of the International Development Research Centre (IDRC) for supporting the work that led to the development of this book. The technical support and guidance of Pascal Sanginga was particularly important in seeing this project through to completion, and the advice of Bill Carman was instrumental in ensuring the final publication of this book.

The authors would like to thank the colleagues and associates who devoted time to peer review chapters of the book, or the book in its entirety: Alex Moiseev, Mine Pabari, Cathrine Mutambirwa, Ed Barrow, Ali Kaka, Leo Niskanen, Katharine Cross, Sue Mainka, Hastings Chikoko, Monica Naggaga and Vanessa Tilstone. Particular thanks to Geoffrey Howard for reviewing the final book and advising on its completion. Gratitude is also due to the Commission on Education and Communication (CEC) for their support in shaping the analysis and their input to key sections of this book.

IUCN would like to thank its members and project partners who contributed to many of the experiences that are presented within. Particular thanks to the governments of Kenya, Tanzania and Uganda for enabling the work in each country. Thanks also to the International Planned Parenthood Federation for its ongoing partnership to understand the linkages between Population Health and the Environment.

In Tanzania, the authors would like to express gratitude to the Rufiji District Council, to YOVODEA (the Youth Volunteers Development Association) for their work in Kibaha district, and to communities from Rufiji who participated in the Project. In Uganda the authors would like to express gratitude to members of TEAN (The Environment and HIV/AIDs Network), RHU (Reproductive Health Uganda), Uganda Population, Health and Environment (PHE) working group, the Manafwa district local government, Bukhofu Women's group, and the Uganda Wildlife Authority in Mt Elgon Conservation area. In Kenya additional thanks are due to the Resource Advocacy Project in Garba Tula.

Regionally, we would like to thank the IGAD Secretariat, and in particular the Division of Agriculture and Environment for their enthusiastic partnership in key areas of work that have informed this book. We are also grateful to the East African Community for their support and partnership.

1

CONSERVATION AS A LIVELIHOOD STRATEGY

Linking policy with practice

Jonathan Davies

The loss of biodiversity and vital ecosystem services presents an important barrier to the achievement of the Millennium Development Goals to reduce poverty, hunger, and disease. Human Development depends on nature and ecosystem services: growing populations, with their associated demand for food, water, fibre and energy, have caused unprecedented changes to ecosystems. While nature has contributed to improving billions of lives, development has simultaneously damaged the ecosystems on which it depends, and this degradation is increasing (MA, 2005).

Although the intimate relationship between the environment and development is well researched, it is often poorly reflected in policies and development planning. While sustainable use of natural resources may receive some consideration, appreciation of how ecosystems function and their importance for sustainable and resilient livelihoods is low. It is vital to raise awareness amongst government decision-makers to motivate and enable them to enact policies and plans that can protect the long-term viability of their development processes (UNDP-UNEP, 2009).

Conserving natural resources requires greater ownership and benefit sharing for local communities and their greater involvement in decision-making. It also requires greater coordination within and between governments, businesses, and international institutions (MA, 2005). Realizing this, conservation and sustainable development actors pay growing attention to communicating with decision-makers in an attempt to influence policy and practice. Many innovative approaches have been followed, to lobby or advise governments, through formal and informal channels of communication, and with widely divergent results. Non-governmental actors increasingly try to influence government decision-making, and in some cases governments are becoming more open to input from such actors, through public meetings or private consultations.

Despite this growth in efforts to influence government decision-making, evidence of the impact is sometimes thin on the ground. Coordination between different groups that are attempting to influence the same processes and decisions is often weak and their complementary roles are not well harnessed. Evaluating and learning from different advocacy efforts is weak and advocacy often lacks strategic direction, or is conducted without adequate insight to the complexities of decision-making processes. Above all, the inherently political nature of policy dialogue is poorly understood, and therefore opportunities to influence decision-making are often wasted.

A number of high quality publications can be found that provide advice on influencing policy. These resources typically focus on one angle of influence: for example, the use of evidence and science, or the mobilization of citizens. What is lacking is a strategic overview of the diverse range of approaches and actors who can engage in different levels of decision-making, from local to national, and at different stages in decision-making processes from setting the agenda to implementing a policy. Such an overview needs to also consider the type of influence different actors can bring to bear, from cold hard science to social and political pressure, and the different types of government decision that may be susceptible to each.

This book sets out to provide such an overview and develop a framework for more strategic planning to influence government decision-making. Each chapter draws on a particular range of experiences in harnessing the influencing power of different stakeholders, from community natural resource managers to government technical advisors, and from locally elected leaders to inter-governmental institutions. Each of these actors has certain areas of legitimacy, privilege and capacity that it can use to influence certain kinds of government decision-making processes. If these actors are also aware of each other's advantages, they may be more able to form more strategic alliances. By understanding not only the role of non-governmental actors in decision-making processes, but also the role of different government actors, it may be possible to create policy partnerships that enable governments to make more effective decisions and simultaneously achieve greater local ownership over their implementation.

Many of the case studies and examples presented in the chapters of this book come from projects implemented by the International Union for Conservation of Nature (IUCN), with support from the International Development Research Council (IDRC). The IUCN is a unique body that comprises over 1,100 institutional members, including governments, government departments and non-governmental organizations. It also comprises six international commissions with over 11,000 expert members.[1] This union provides a great diversity of opportunities for engaging in decision-making, by drawing on science as well as local expertise, and convening dialogue between very different institutions and representatives from local to global levels. The book draws on this diversity to explore the wide range of opportunities for influencing government decision-making processes.

Conservation as a livelihood strategy

Natural resources form the backbone of many developing country economies, and rural households in Eastern Africa are heavily reliant on the natural environment for their livelihoods and income. A healthy and productive environment is therefore vital for human wellbeing and for pro-poor economic development. Ecosystems provide a range of life-supporting services, including the provision of food, water and fuel, which underpin rural and urban livelihoods. Furthermore, they play a vital role in regulating climate and other environmental shocks and are crucial for the management and mitigation of risk.

Sustainable management and use of natural resources and ecosystem services is well recognized as an important ingredient for sustained economic development and improvements in human welfare, and therefore is necessary for achieving the Millennium Development Goals (UNDP-UNEP, 2009). Environmental assets, such as soil, water and biodiversity, yield income, offer safety nets for the poor, maintain public health and drive economic growth. Yet environmental sustainability goals are often seen as distinct from, and sometime in conflict with, development goals (UNDP, 2008).

This false dichotomy, between environment and development, contributes to the environmental challenges that the world currently faces, including mass species extinction in the face of human activities, and human-induced climate change. Efforts have been made, particularly in the two decades since the

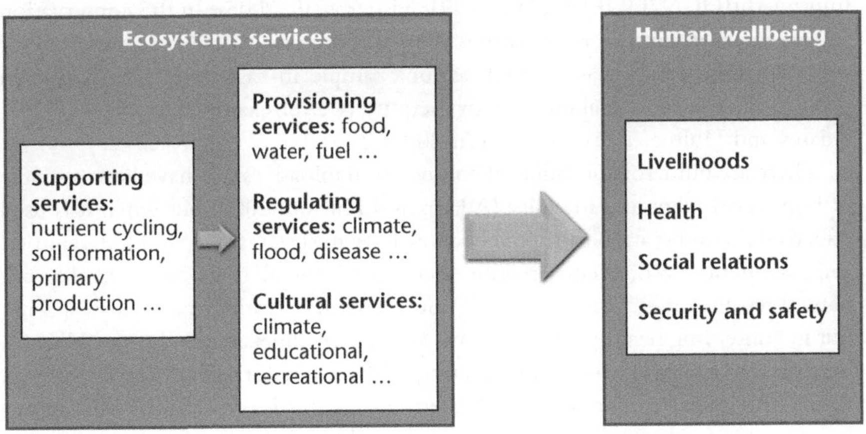

FIGURE 1.1 Linkages between ecosystem services and human well-being (Millennium Ecosystem Assessment, 2005)

Box 1.1 ***Natural Resource Reliance in Eastern Africa***

- The contribution of agriculture to GDP of the United Republic of Tanzania and to Uganda is 46 per cent and 45 per cent respectively, while the percentage of the workforce employed in agriculture is 81 per cent in both countries (Bishop-Sambrook, 2005).
- The percentage of the population without electricity access, and therefore reliant on natural biomass fuel sources, is 85 per cent in Kenya and 91 per cent in Uganda (Legros *et al.*, 2009).

Brundtland Commission and the Rio Earth Summit, to mainstream environmental concerns in development, but these efforts have a poor track record. There remains a tendency to treat the environment as an institutional and economic externality, and there is still a widespread dearth of data, skills, institutional capacity and political will to achieve meaningful change. Integrating environmental and development concerns can enable improvements in productivity, resilience and adaptability of livelihoods and economies. However, it has become clear that such integration requires political and institutional change as much as a technical development (Dalal-Clayton and Bass, 2009).

Development is considered to be sustainable if it improves the quality of human life without exceeding the carrying capacity of supporting ecosystems. Sustainability is usually thought of has having three inter-related components: economic, environmental and social. These three "pillars" need to be addressed simultaneously if development is to be sustainable, and people's livelihoods are to be improved (IUCN/UNEP/WWF, 1991; Fisher *et al.*, 2005). In the conservation sector, interventions have sometimes proven to be unsustainable because one or other of these pillars was ignored, as for example in cases where conservation projects have led to social injustice or inequity (see for example, McNeely, 1993; Adams and Hulme, 2001; Neumann, 2004).

There are numerous accounts of conservation initiatives that have been a source of impoverishment and inequity (Adams and Hutton, 2007) but when it is used effectively, conservation can actively contribute to sustainability and to livelihood improvements. When conservation becomes integral to people's livelihoods natural resource users may see benefit not simply in operating at carrying capacity, but in conserving healthy environments and species diversity that bring additional benefits, for example, through marketing of diverse natural products, or through forms of nature-based tourism. In Eastern Africa this is of particular importance because the region is characterized by relatively high levels of poverty, but has great environmental wealth and many opportunities for conservation-related enterprises.

Mainstreaming conservation in policy and planning

Policies are frequently found to undermine sustainable livelihood development and sustainable natural resource use. An example that was particularly widespread in post-colonial Africa, and which has had important ramifications for sustainable natural resource use, is the policy of nationalization of land, which has shifted power and authority and weakened some traditional leadership, thereby undermining local governance of natural resources (Toulmin and Quan, 2000). However, simply labelling policies as "failed" may be unhelpful and the objective of a given policy may simply be at odds with local requirements, while satisfying a different goal at a national level: for example, policies of agricultural intensification and industrialization that are driven by a pursuit of foreign exchange. Nevertheless, as attention shifts to pro-poor development and adaptive capacities of rural populations, and as environmental concerns move into the limelight, it is important to examine whether the current policy environment is conducive to sustainable rural development.

The field of international development has changed radically over the past half century and this has had profound implications for the way governments develop policies, how they integrate environmental concerns in policy, and how non-governmental actors engage in the process of environmental mainstreaming and sustainability. Put simplistically, the trend since the 1950s has been from macro-level economic development, revolving around promotion of investment and project planning, through a more growth-oriented "trickle down" and import substitution approach in the 1960s, to more localized social development focusing on basic needs and redistribution, and eventually to the notion of "development as freedom" (Sen, 1999). In parallel there has been a steady increase in engagement of non-governmental actors in policy and planning processes.

During the 1980s, questions of sustainability came to the fore, leading to the United Nations in 1983 convening the World Commission on Environment and Development (WCED). Also known as the Brundtland Commission, this initiative addressed the growing concern about the accelerating deterioration of the human environment and natural resources and the consequences of that deterioration for economic and social development. The report of the Commission (World Commission on Environment and Development, 1987), also called the Brundtland Report, places emphasis on the importance of sustainable development and the change of politics needed for achieving that. The definition of sustainable development used in the report is still widely cited: "Sustainable development is development that meets the needs of the present without compromising the ability of future generations to meet their own needs." The United Nations Conference on Environment and Development in Rio de Janeiro in 1992, also referred to as the Earth Summit, followed on from the Brundtland report. This conference led to a number of high-level international agreements, including the adoption of the Convention on Biological Diversity.

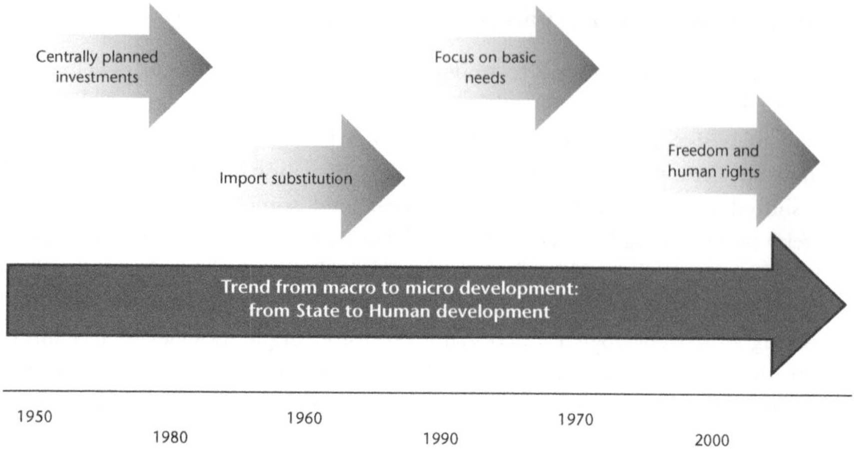

FIGURE 1.2 The trend towards greater inclusivity in development planning

The two decades since the Earth Summit have seen increasing emphasis on the institutional and social aspects of development, including poverty reduction, building capable states, good governance, conflict prevention and resolution (Sagasti and Alcalde, 1999). The 1990s also saw the concept of Sustainable Human Development put forward by UNDP to integrate economic growth, social development and environmental sustainability. At the same time there has been growing attention to Rights Based Approaches and empowerment as both a means and a goal in development processes (Sen, 1999).

The current global development situation shows mixed results and major challenges in attributing successful outcomes to shifting development paradigms. Major success stories include countries such as Turkey, Brazil, India and China, but many less positive outcomes can be cited in the developing world, with increased global poverty, greater disparity between rich and poor nations and between citizens within nations, and severe environmental stress (UNDP, 1994). However, out of the success and failures of development, a number of global lessons are being learned, amongst which is the importance of adequately integrating environmental considerations in the design of development strategies and policies to ensure sustainability.

As sustainable development has evolved, emerging lessons have begun to influence planning and policy at different levels. There is now increasing understanding of the importance of building capacity to acquire, generate and use knowledge, including indigenous or traditional knowledge, for improving material living standards (World Bank, 1998). Greater attention is given to institutional issues, for example, strengthening institutions associated with democratic governance or for channelling social demands towards centres of power, and institutions associated with markets and competition (Sagasti and Alcalde, 1999).

Policy dialogue for environmental mainstreaming

With the emergence of more people-centred development in recent years, there has been a steady increase in the number of positive experiences that can be drawn upon to inform further success, yet many development agents have become frustrated in their efforts to scale up localized good practice, often achieved through an intense spell of NGO activity. Policy failures and disincentives are often seen to act as stumbling blocks to spontaneous uptake of innovation, and the result is an increasing interest in policy advocacy, or policy dialogue, to try and rectify these constraints. In parallel there is a growing interest in citizen participation in policy-making as an empowering and educative process: as a means to strengthening rights and responsibilities in developing countries.

In recent years, Civil Society has won a growing number of opportunities to speak on behalf of citizens in national and international fora on a wide range of topics. At times Civil Society has been relegated to presenting their case at parallel or unofficial sessions of formal meetings, but increasingly they are claiming the right to take part in formal decision-making processes with official delegate status (Holmén, 2002).

Civil Society typically engages in policy dialogue and advocacy in order to achieve greater impact and to move beyond short-term service delivery. Not all governments are open to input from Civil Society and some countries still suppress such voices, or are at least resistant to CSO engagement, but opportunities are growing. However, despite evidence that the political environment in some countries has opened up to Civil Society, many groups fail to take advantage of the opportunity. This is partly due to continuing constraints and obstacles, and partly due to lack of capacity and know how on behalf of Civil Society Organizations. Civil Society faces particular challenges in cooperating effectively, in building alliances, in developing legitimacy or credibility, and in wielding effective evidence. Research can play a strong role in policy development and thereby impact positively on sustainable development, but understanding of how to use research systematically and to greatest effect is weak (Young, 2005; Court *et al.*, 2006).

Nevertheless, a growing number of non-governmental agencies engage in policy dialogue or advocacy for different reasons and with different levels of influence and effectiveness. Agencies may engage in policy dialogue to enable the scaling up of good practices, for example, by removing bottlenecks or promoting public investment and attention to a given issue. Policy dialogue may be pursued to demand regulation or controls, for example, in the case of environmental pollution. Policy dialogue may also be used as a vehicle for empowerment of poor communities, through engagement of citizens in the process as much as through any particular policy outcome (Kutugeka and Roba, Chapter 2 this volume).

Worldwide attention to environmental challenges is growing, bringing ever more actors into policy debates, with emerging alliances and changing roles for science, government and Civil Society. This creates both opportunities and constraints for moving towards a more enabling policy environment for sustainable development, as more avenues for dialogue are created. At the same time new alliances and power bases are emerging, for example, with the disproportionate power of Aid Agencies in relation to comparatively weak Civil Society in Eastern Africa. This contributes in some cases to growing scepticism of purely technical advice in policy-making processes and a growing interest in and awareness of more open, public consultations (Holmes and Scoones, 2000).

Models and interpretations of policy influence that are used in development work have been accused of being dominated by experience from industrialized societies. These models make unrealistic assumptions about participation of citizens in policy processes, and about the rationality and predictability of institutional arrangements and policy processes. The reality in developing countries is quite different where it can be particularly challenging to bring new arguments and good science to bear on decision-making processes. In many developing countries democratic institutions and practices are weak and policy-makers lack autonomy and may be more susceptible to influence from personal relationships. At the same time, policy-oriented research in developing countries may be weak and intermediary institutions to translate research into policy may not exist. As a result, policy-makers often lack confidence in researchers and their demand for research is low (Carden, 2009).

In countries where trust is weak, whether between people and government or between different interest groups, inclusive policy-making processes are particularly important. Consultative processes take on yet greater importance in relation to policies where different interest groups are highly active, with entrenched positions and where the stakes are high. In many developing countries, inclusive, consultative or participatory policy processes are favoured by many actors outside government, because trust towards government is weak and the State is associated with corruption, failure to deliver basic services, and misgovernance (Holmes and Scoones, 2000).

Consultation can take many different forms: governments may deliberately include people in consultations; community groups may force their issues onto the policy agenda; or political representatives may be used to influence policy (Gaventa and Robinson, 1999). Where debate, conflict and differences of opinion are prevalent it is more, rather than less, important to have a consultative process and it is important to reach consensus and compromise (Holmes and Scoones, 2000).

Participation in policy-making is rarely genuinely democratic and usually only includes a minority of the people that a policy affects, which raises important questions of who is to be consulted (Lapintie, 1998). There is always a risk that consultative processes will favour a select group of individuals in a society, such as the wealthy, those with higher status, or people with a particular motivation that may not be representative of the wider community. To overcome these risks, third parties often intervene in identifying participants for consultation, and this is particularly useful for enabling minorities and other marginalized groups to engage in policy processes. Even then, it is hard to ensure equitable representation of different groups in a society, such as women (Holmes and Scoones, 2000).

Understanding policy, planning and government decision-making

Although there has been a growth in policy dialogue between non-governmental actors and government, and at different levels from local to national and to international, there are many divergent opinions of what the term "Policy" means. It is useful to recognize that influencing policy is a means to an end, and it is important to keep the end in mind, rather than get distracted by definitions. In the context of environmental mainstreaming, the objective is often to integrate environmental considerations in decision-making at each stage of the decision-making process. This could apply to decisions made by central government, for example in creating legal frameworks or allocating financial resources, but it could also apply to resource allocation at local level, or to a number of other decision-making processes.

Some definitions of policy are so broad as to encompass norms and culture, which blurs the distinction between policy and practice. Using such a broad definition, localized good practice may simply be policy at a local level. Influencing "policy" can therefore entail influencing a wide array of decision-making processes and this requires a range of different approaches and the participation of diverse stakeholders. The chapters in this book use the term "policy" in a broad sense and present examples of influencing a wide range of government decision-making processes. There is a common focus on policy of government, but at the same time it is recognized that many advocates also attempt to influence the policies of other NGOs or International Organizations.

In relation to government policy, there seems to be consensus at least that policies are guiding principles or lines of argument that are designed to rationalize a course of action and influence the decisions and actions of an institution such as a National Government.

> Although synonymous with law, rule, statute, edict, and regulation, public policy is the term of preference today probably because it conveys more of an impression of flexibility and compassion than the other terms. But citizens . . . should always be aware that policy and police have common origins. Both come from "polis" and "polity", which refer to the political community itself and to the 'monopoly of legal coercion' by which government itself had been defined.
>
> (Lowi, 1998: foreword in Tatalovich and Daynes, 1998)

Box 1.2 *Definitions of policy*

- A line of argument rationalizing the course of action of a government (www.thefreedictionary.com/policy).
- Any standard, statement or procedure of general applicability adopted by an institution – something that guides the institution's operation (www.ncsu.edu/policies/definitions.php).
- A principle or rule to guide decisions and achieve rational outcome (http://en.wikipedia.org/wiki/Policy).
- A plan of action, usually based on certain principles, decided on by a body or individual (*Chambers 21st Century Dictionary*).
- A principle or set of principles on which to base decisions (*Chambers 21st Century Dictionary*).
- A course of conduct to be followed (*Chambers 21st Century Dictionary*).
- A set of decisions that are oriented towards a long-term purpose or to a particular problem (www.fao.org/wairdocs/ILRI/x5499E/x5499e03.htm).
- Prudence; acumen; wisdom (*Chambers 21st Century Dictionary*).

Policies may not always have as much impact as government (or others) would like, and policy is often seen as trying to catch up with practice. Nevertheless, policy can play an important role in determining government expenditure and priorities, creating legal frameworks, and in legitimizing changes on the ground.

It might often be assumed that public policy stems, one way or another, from politics, yet Lowi (1998) contends that causality is the reverse: politics are shaped by and emerge from policy. This observation may have derived from the USA at a time when party politics were relatively ideologically driven. However, the observation is important for Eastern Africa given the role that Civil Society and citizens are increasingly playing in the policy arena, and the importance of policy processes in underpinning democratization (Schneider and Ingram, 1997). If policy is about power, then we have to be aware of how power relations can change through policy engagement and of the growing influence of citizens in Eastern Africa.

Clarifying the policy process

Many efforts have been made to rationalize the policy-making process, usually depicting a cycle with a varying number of components. Kingdon (1984) identifies four important steps, as illustrated in Figure 1.3, focusing on the initial stage of

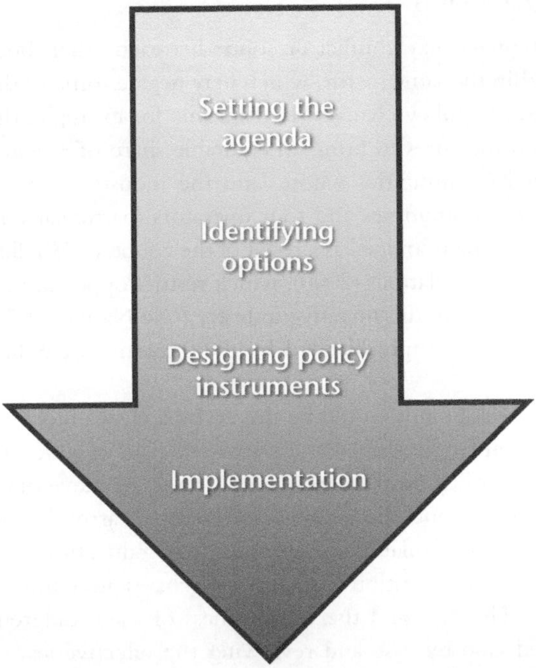

FIGURE 1.3 Steps in the policy process (Kingdon, 1984)

identifying the issue – which may mean identification of a problem or an opportunity – the weighing up of options, official endorsement of one or more of those options and finally the implementation of the policy. Other authors have included different sub-steps such as evaluation of the policy, consulting and coordinating or distinguishing between identifying the policy option and writing the policy (Lasswell, 1977; Young and Quinn, 2002; Pollard and Court, 2005).

Some analysts consider government decision-making as a cyclical and iterative process. Typical components in the cycle include design, implementation, evaluation and diagnosis of problems. Some cycles elaborate further on the design stage, including steps such as analyzing the options, developing instruments, consulting and coordinating. Yet government policy often does not follow such a rational or systematic cycle and these cycles should not be viewed rigidly as the way that decisions necessarily are made: in practice steps are ongoing and overlap. The cycles are a conceptualization of the nominal steps that decision-makers may go through, and they help in identifying entry points for influencing from the outside, or for entering into the process (Dalal-Clayton and Bass, 2009).

Inconsistencies in policy-making

Within a country, different policies may conflict or contradict each other, both between sectors and even within the same sector, which may negate some of the benefit of bringing about positive policy changes. In Tanzania for example, the 1998 Wildlife Policy proposed measures to bring an equitable share of revenue from tourist operators to rural communities whose land the industry uses. In practice, relationships between communities and tour operators (particularly in the hunting tourism sub-sector) remain strained as a result of the country's Wildlife Conservation (Tourist Hunting) Regulations (2000), which restrict opportunities for communities to reap the benefits of this lucrative industry (Ole Nasha, 2007). Decision-making clearly cannot be simply dictated by the existence of policy documents.

In an ideal world, policy development would be driven by a clear objective, and a Policy Strategy would then be developed as a systematic plan of action to achieve the policy objective. Policy instruments would then be developed to provide the means of implementing the strategy, including "carrots" (e.g. subsidies), "sticks" (e.g. emission regulations), sermons (e.g. education and voluntary instruments), and direct intervention (e.g. direct purchases, investments and the creation of markets). The cost and the effectiveness of these different instruments would be assessed case by case and reviewing the effectiveness of these instruments would be an important part of the policy-making process. Policy-making would not necessarily demand a single objective or a single instrument, and multiple objectives and instruments are often found, although this can lead to, or may stem from, a conflict of interests (Torjman, 2005).

In reality, such a systematic approach to policy-making is often not adhered to, and the distinction between policy objective and policy instrument may be blurred. Instead, interest groups pursue their own agendas: either to secure their chosen objective, or to ensure a particular instrument is given priority.

> Policies in most governments, most of the time, are the outcomes of all the bargains and compromises, beliefs and aspirations, and cross-purposes and double meanings of ordinary governmental decision-making. This is why it is usually a mistake to adopt a model that imagines policy-making as a rational, orderly, or unitary and linear progression from problem to decision and solution.
>
> (Carden, 2009)

Government planning processes at different levels may be guided by specific policies and legal frameworks, but often such processes can be arbitrary and reflect unwritten "policies", or political (or other) agendas. Government planning is carried out to make decisions about the problems and issues to be addressed and the order of priority, usually in relation to specific, limited available public funds.

This can require consensus building, although such processes are often imperfect and frequently are not consultative beyond a narrow inner-circle of planners. However, participation should focus not simply on production of a plan, but should be designed to create dialogue space between multiple stakeholders with divergent goals and understanding. Planning processes are required to reconcile opposing and conflicting demands and the process is therefore critical.

Influencing decision-making

When decision-making processes are complex, or when they lack transparency, there are many challenges for non-governmental actors to use their resources, profile and skills to greatest effect. These actors have to decide whether to engage with government or with non-government authorities, with ministries of environment, finance or other sectors, whether to work upstream on policy development or downstream on investments and implementation, and whether to mainstream environmental issues or to create special processes (Dalal-Clayton and Bass, 2009).

The role of citizens in environmental policy dialogue has received particular interest in recent years, benefiting from the emergence of rights-based approaches to development. In the fields of conservation and the environment, ethical and moral considerations take great prominence and this often demands a shift from purely rationalistic dialogue towards more inclusive and consultative approaches (Holmes and Scoones, 2000). Citizen participation in decision-making has been boosted by processes of decentralization in many countries, since decentralization has created more opportunities to build relationships between citizens and government and to foster dialogue. Decentralization has contributed to improved understanding of political processes, representation and the nature of policies plans and budgets. It has also contributed to improving the skills of political leaders in policy-making and planning, greater equality and de-concentration of decision-making power and has increased accountability and responsiveness of government (Turner and Hulme, 1997).

Opening consultation to a wide range of stakeholders cannot be assumed to guarantee that environmental concerns will be mainstreamed. Stakeholders pursue different interests and not all of these can be assumed to be environmentally or socially sensitive, or indeed in the interest of sustainable economic development in general. Effectively influencing decision-making will often require understanding the motives and rationale of competing stakeholders in order to identify appropriate arguments or relevant compromises to reconcile conflicting interests (UNDP-UNEP, 2009). However, citizen participation is important for empowering local communities with regard to powerful outside interests and technical experts (Kutugeka and Roba, Chapter 2, this volume). This creates opportunities for harnessing the local skills and knowledge that are vital for the conservation of natural resources (Kamagenge, Chapter 3, this volume).

Who represents who?

Given that participation in policy-making cannot realistically include everybody concerned, the question of who represents whose interest becomes particularly important. Community delegates or spokespeople are often credited with a representational capacity that they do not possess, and frequently they are unable to speak for an entire community (Hoyes *et al.*, 1993, cited in Holmes and Scoones, 2000). This can be the case for spokespeople identified by outsiders, such as NGOs, but may also apply to spokespeople appointed by a community, such as elders. In many cases, local 'representatives' are eager to assume the role of representation, but either they do not represent everybody, or they are unable to represent on every issue (ESRC, 1998).

Various approaches are used to overcome these issues of representation, including working with democratically elected representatives, and working with communities to ensure that they fairly appoint delegates to represent their interest. Other ways to improve the representation of a wider cross section of a community include multiple stakeholder participation, informal representation through interest groups (Rossi, 1997), and inclusion of NGOs and CBOs in policy processes (Zazueta, 1995), although each approach has its costs and benefits, and its champions and its detractors. A number of innovative approaches to working with representatives are presented in this volume in Chapters 4 (Wandago) and 6 (Nakangu Bugembe).

NGOs often go to great lengths to identify community representatives, sometimes setting up parallel processes to the constitutional process that exists in most East African States. However, in many countries, and in many constituencies, elected representatives are in place with a clear role and mandate to raise the concerns of their constituents in policy fora. Elected representatives may not always be reliable, and cannot be considered to represent the interests of everybody in their constituency, but some parliamentarians are very active in the policy arena and are prepared to represent the interests of their constituents against outside interests. Successful participatory policy-making often depends on integrating participatory processes with political representation, and in countries where local representatives are democratically elected, and where communities understand the democratic process, those leaders are legitimate appointed representatives and can be considered to speak for their constituencies (Selman, 1998).

It is challenging to identify the right people to consult in a policy process, and it can be hard to ensure that the right people are interested in such consultations. Citizens often lack motivation to engage in policy processes, and this may reflect a lack of understanding of policy, or it may reflect disillusionment with a process that is perceived as unrepresentative and opaque. Indeed, this lack of motivation may be the very reason why consultation is required (Eden, 1998). Building trust in a process may therefore go hand in hand with developing participation, since motivation may stem from belief in a successful outcome.

Motivating people to engage in a policy process can help to ensure that a policy is guided to a successful implementation, but it is also a means of raising awareness of constitutional processes and citizenship. It is therefore inextricably linked with building trust in government, and is clearly therefore a two-way process in which people have to be empowered and government institutions have to become more accountable. This is no small feat, but to some extent it is possible to foster trust by developing dialogue between citizens and government and pursuing a more protracted deliberative process, in which decision-makers and communities engage in open discussion and collaboration, rather than focusing on short sharp advocacy actions (Healey, 1997).

Whose expertise counts?

The significance of science and technical expertise in determining policy is sometimes contested, although much money is invested in research to influence policy. The nature of the policy under discussion may determine the extent to which government includes or excludes expert advice and opinion. However, in many quarters trust in expert advice has become tarnished by the perception that science is socially constructed and elitist (Keeley and Scoones, 1999) and even politically captured (Eden, 1996). Key decision-makers at different stages of policy development or government decision-making in developing countries may lack the capacity or inclination to use science, or may not have access to the science they require. They may view science as externally driven, conducted and controlled by actors outside their country such as the World Bank. As a result science often fails to influence policy (O'Neil, 2009: Foreword in Carden, 2009). Furthermore, an over-reliance on science in policy discourse can be disempowering for communities if it fosters a sense of exclusion (Sanderson, 1999). An ideal situation is for non-scientists to be supported to conduct their own research and translate their own knowledge into "expert" advice, making the process much more participatory (ESRC, 1998).

An alternative, or perhaps a complementary, approach is to promote the voice of local experts and the importance of local or indigenous knowledge. Participatory approaches are often favoured in rural development, based on the assumption rural communities have rich knowledge about their natural resources and about how to manage them effectively, and with appropriate support they can become important agents of change (Wyckoff-Baird *et al.*, 2000). Since decision-making is such a complex and highly political process, based on negotiations between different lobbyists, it is easy for some groups to be marginalized from the process, perhaps due to their political exclusion or due to inadequate awareness of citizenship and constitutional rights. Often these excluded groups are amongst the poorest in their country and therefore pro-poor development requires an investment in enabling them to influence policy and to safeguard their interests. Nevertheless, it is important to determine where local knowledge is strong and where it is weak, and to avoid over-emphasizing the relevance of local knowledge in all discourse (Thrupp, *et al.*, 1994). Holmes and Scoones (2000) found that where environmental policy has been effectively influenced, consultative approaches have combined science with local knowledge, for example, where local communities are assisted to adopt expert advice in their advocacy.

One area where expert knowledge can play a significant role in environmental policy is in making the economic case for investment in the environment. National statistics rarely capture the asset value of nature or the economic value of ecosystem services, and as a result the real environmental costs and benefits of different policy options are easily overlooked. To counteract this it is necessary to invest in economic accounting so that policy-makers understand the real cost of a particular investment or policy trajectory. It is also important to support citizens to engage effectively with local governments and politicians in order to present this case (Adams and Jeanrenaud, 2008). Chapters 5 (Ogali) and 7 (Manzano Baena) of this volume discuss the challenges and opportunities of linking science and research with government decision-making. Chapter 8 (Hesselink and Zeidler) also explores a wide range of innovative forms of communication that have been particularly effective at improving the influence of research on decision-making.

Making the linkages between policy and practice

The apparent increase in non-governmental engagement in policy and planning processes appears to reflect a combination of factors, including deeper appreciation of the importance of those processes, growing awareness of the significance of strengthening community participation, and increasing emphasis on empowerment as a development goal. Certainly there are many diverse initiatives, within Eastern Africa as well as in other developing regions, to influence government processes, yet the impact of these initiatives is often difficult to determine. This is not necessarily an indictment of those initiatives, since there are tremendous challenges in tracing real impact, on the environment or on economic development, back to individual contributions to decision-making. Actors that engage in policy dialogue or advocacy face a continual struggle to justify their actions to results-oriented donors, and to attribute impacts and outcomes to specific actions.

A number of the case studies that are presented in this book have been contributed from project activities implemented by the International Union for Conservation of Nature (IUCN), in Eastern Africa. Like many of the other case studies cited here and elsewhere, these interventions make significant assumptions over how policy dialogue will lead to real changes in the environment or in people's lives. It is incumbent upon actors engaging in policy dialogue to clarify the steps they are taking in order to know where they fit into the causal chain that links knowledge and action. This will improve accountability and learning and enable a deeper reflection on the many assumptions that are made.

Policy dialogue is carried out to contribute to sustainable development, and improved livelihoods and biodiversity conservation in the long-term. Part of the goal may also be to empower people through engaging in the process, although empowerment is itself both of intrinsic and instrumental value: both a means to, and a component of, sustainable livelihoods. To contribute to this goal, non-governmental actors often focus on raising awareness in government and on increasing and improving the engagement of communities in policy-making processes. However, there is a gulf between knowledge and participation and the ultimate goal, and the impact pathway needs to be carefully considered, and influence may need to be brought to bear at a number of levels.

Within the broad theory of change that underpins much policy work, whether implicitly or tacitly, is the assumption that improved participation and improved knowledge and awareness amongst different stakeholders will lead to the creation of new policies, the adjustment of existing policies, and the implementation of policies. Such outcomes are not guaranteed, but it is within the power of non-governmental agencies to influence at this higher level: to participate in policy formulation processes, to provide feedback on the shortcomings of policy, or to support government to implement policies.

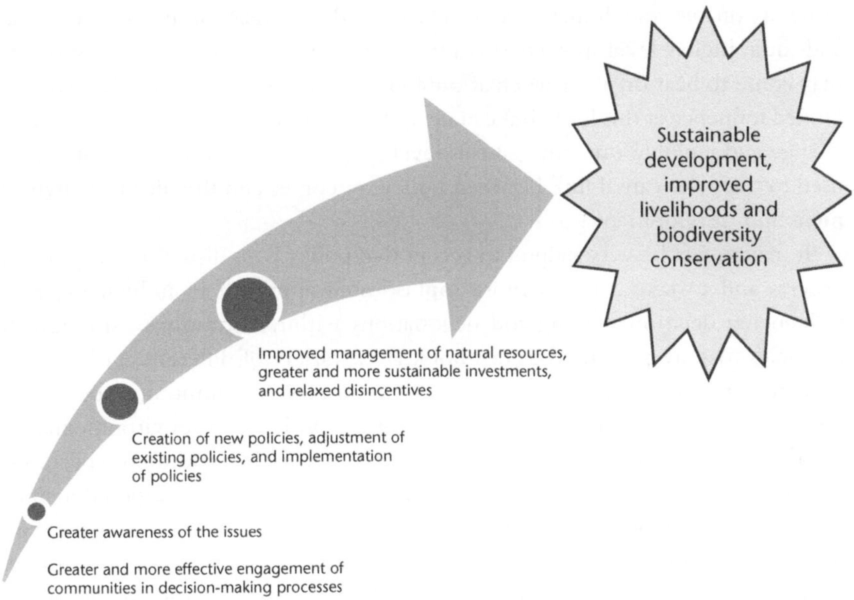

FIGURE 1.4 Generalized Theory of Change

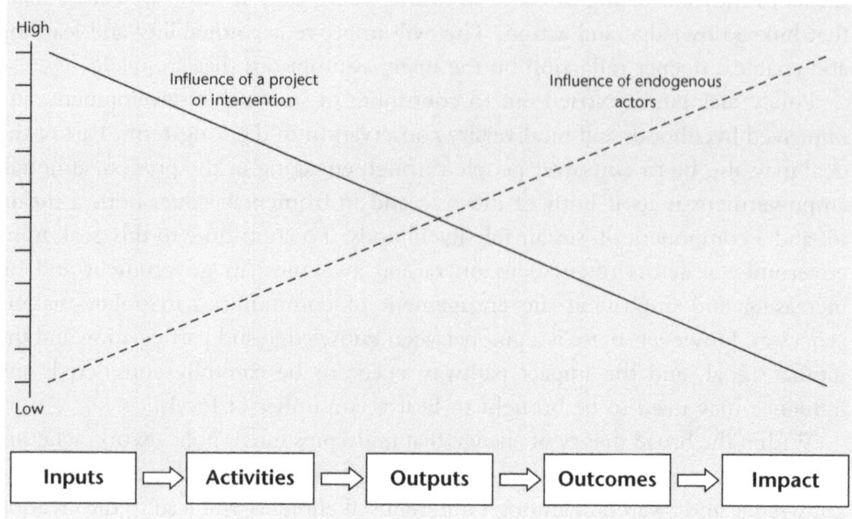

FIGURE 1.5 Relative influence along the results chain (Smutylo, 2001)

A further assumption is made, that with a conducive policy environment, the management of natural resources will improve, investments will be greater and more sustainable, and disincentives will be relaxed. The relationship between policy and these higher level impacts is complex, since multiple policies and processes may come to bear on any particular outcome. An individual institution may have limited influence at this level, but can nevertheless monitor linkages between policy change and tangible outcomes at this level. Monitoring and research can also be used to trace the causal link between policy outcomes and the ultimate environment and development goals.

It may in fact be very helpful to accept that policy is an ill-defined and chaotic process and to take a much more sophisticated approach to influencing it: to influencing decision-making and negotiations within institutions. Amongst the examples that are presented in this book are a variety of different approaches to influencing policy and practice. There are examples of communities persuading local government to allocate resources (money and land) for environmental and sustainable development activities. There are examples of communities becoming empowered through engagement in policy dialogue. Other examples document the development of scientific and economic arguments, combined with community testimony and wielded by different actors in dialogue with decision-makers in government planning departments. There are examples of engaging with elected and un-elected representatives, such as councillors and parliamentarians.

Measuring and attributing change

The Theory of Change that has been outlined in this book and that has informed much of the work presented here illustrates one of the critical challenges to effectively influencing policy: that of knowing what impact your contribution makes. Monitoring the effectiveness of policy influence is critical both for accountability, for example, to allies or to a donor, but also to ensure that lessons are learned and that influence can be steadily improved. However, it is unlikely that one intervention will work alone to influence policy, which leads to a challenge of attributing a particular outcome to a particular intervention. Furthermore, since changing policy is usually a means to an end with a long impact pathway, it is even more challenging to determine the role that policy influence has in terms of real impact: for example, on lives or on the environment.

Many agencies work at several levels in the results chain, for example, starting by improving practice to have a direct impact on people and the environment, and then gathering lessons to feed into policy to ensure that local lessons are (a) sustained and (b) scaled up. Measuring the impact of their efforts to influence policy would therefore require monitoring how well their first intervention is sustained, over a significant period of time, and also monitoring the degree of scale-up nation-wide. Such levels of monitoring are unrealistic for most project-based interventions. Furthermore, even if impacts could be measured, there could still be doubts as to the real causal factors, since many other interventions could be influencing adoption of a particular practice on the ground.

Clearly many assumptions are made when influencing policy and it is important to be realistic and transparent about these assumptions, but also to learn as much as possible about the effectiveness of interventions. It is often assumed, for example, that greater participation of decision-makers in dialogue with other stakeholders will lead to more sensitive and appropriate policies. It is also often assumed that if sound science is used to identify policy options, policies will be improved. These assumptions, as the chapters in this book illustrate, do not always hold true, since there are many competing interests and many different decision-makers and policy influencers. The Theory of Change is helpful in this regard, since it allows actors to clarify how their work contributes to their overall goal and to assess the effectiveness of their influence.

To overcome the challenges of monitoring and attributing impact, it can be beneficial to examine the outcomes: that is, consequences of the intervention that are expected to contribute towards the ultimate impact. The Outcome Mapping approach can be used to visualize the desired future situation – the improved livelihoods and environment – and then to identify the steps towards that future scenario to which a given intervention can contribute. Central to this approach is the understanding that a single intervention cannot instigate the desired changes in isolation, but can contribute to changes in attitude and behaviour among other actors who in turn can bring about the desired change. Outcome Mapping is therefore an approach to monitoring social change, measured in terms of changes in behaviour of "boundary partners" (Earl *et al.*, 2001; Smutylo, 2005). Boundary partners are "the individuals, groups, and organizations with whom the program interacts directly and with whom the program anticipates opportunities for influence". The Outcome Mapping approach assumes that these actors have more control over the desired changes that a programme aims to bring about, while the programme itself plays a facilitating role, for example introducing new ideas, opportunities and resources.

Outcome Mapping enables a programme to focus on where it makes its contribution to outcomes, and makes certain assumptions about these outcomes in turn contributing to the desired results or impacts. It is particularly appropriate in programmes that seek to empower communities, devolve power and contribute to sustainable changes in the future (Earl *et al.*, 2001). The examples presented in this book illustrate a variety of ways that diverse actors have engaged in policy processes. Some of these are direct allies, while others may be targets of advocacy or information. These can all be considered "endogenous actors" in the interventions of a given actor, and Outcome Mapping provides a way to enable the principal actor to observe changes wrought by their activities of persuasion and communication. Outcome Mapping was therefore used as a monitoring approach for many of the interventions presented here, to track interventions that explicitly harness the skills and credentials of diverse actors in alliances.

Brief outline of the book

The examples that are documented in this book come from a wide range of stakeholders, particularly in Tanzania, Uganda and Kenya. Examples are taken from fishing communities on the coast of Tanzania, pastoral communities in Northern Kenya, and farmers in Uganda, as well as community managers of forest and water resources throughout Eastern Africa. They include the experiences of government technocrats, members of parliament and representatives from inter-governmental institutions. They are also taken from a wide range of fields, including HIV/AIDS and the environment, marine natural resources management, transboundary water management, dryland natural resources management, and community forest management.

The project that led to the production of this book focused on making the linkages between policy and practice, and the chapters highlight a variety of ways in which the inter-linkages have been tested. To an extent, this book also blurs the distinction between policy and practice, focusing on government and community decision-making with different degrees of formality. What some people refer to as policy may sometimes simply be the way things are done, and often nothing is written down or formalized until it becomes evident that the way things are done is inadequate or needs regulation. Once a policy is written down, its impact may be negligible, while in some instances, policy is written down after a practice has become the norm. This book is intended to contribute to the understanding of these complex decision-making processes and some of the variety of ways that decision-making can be influenced by different actors. By working in partnership, it is possible to mobilize influence at different levels by different stakeholders using different approaches. This clearly highlights the importance of forming alliances and networks to strengthen influence.

Chapter 2 presents experiences from Eastern Africa in supporting communities to influence government decision-making, particularly at the local level. It illustrates how communities can use their experiences to assist government to implement its policies and how communities can mobilize government support for their own priorities. Chapter 3 builds on this by illustrating the importance of networking as a practical tool for enhancing community empowerment. The chapter provides examples of bringing community experiences and knowledge into government decision-making at local and also at higher levels.

Chapter 4 moves beyond community members to discuss the role that leaders and representatives can play in bringing community expertise and opinions into policy dialogue. The chapter examines official government structures for representation as well as the role of traditional leadership, and explores how effectively community voices are expressed through their different representatives.

Chapter 5 explores the role of technical advisors within government – the group sometimes referred to pejoratively as "technocrats" – in defining and shaping policies. The chapter examines efforts that have been made to influence this influential group, and also examines their influence at different stages in the policy process. Chapters 4 and 5 therefore present experience of working with two central parties in policy-making: the legislature and the executive. Both are important to understand if an actor wants to influence policy, but neither is likely to hold all the answers.

Chapter 6 considers the role of Intergovernmental Bodies (IGBs) in influencing policy and shaping policy agendas. Eastern African States are members of numerous IGBs and there are specific roles these institutions may play in shaping national policy. The influence of these institutions may be limited, or confined to certain areas, but a number of experiences are presented of collaborating with IGBs to promote dialogue on certain areas of concern.

Chapter 7 outlines the role of science in influencing policy, and presents examples of where science has been influential. It also examines the theory of science–policy interaction, which highlights some of the inconsistencies that arise between the theory and the practice. Several chapters in this book highlight the political nature of decision-making and the disregard that can be shown towards scientific insights. Chapter 7 reminds us of the importance of ensuring that science continues to be used to avoid policy and investment failures.

Chapter 8 makes a powerful argument for greatly improving the capacity to communicate effectively. In contrast to Chapter 7, Chapter 8 maintains that it is not what you say so much as how you say it that is taken most seriously. A great array of examples of different communications approaches is presented and would-be policy advisors are advised to think "out-of-the-box" when developing their communications, to ensure that their messages really hit home in the simplest and most affordable ways.

The final chapter of this book attempts to draw an overarching analysis of these chapters and present a framework for more effective planning for policy engagement. Chapter 9 presents a framework for more strategic dialogue with government at different levels to apply influence at the right place and time, and with the right actors. The chapter emphasizes that there are many ways to influence decision-making at different levels of government and that having the maximum influence requires taking a flexible approach to who is influenced, who does the influencing, and what we actually mean by "policy". Understanding the unique policy and political environment in each country is a vital starting point, as is identifying the right allies. The key message of this book is that there is a great deal that actors can do to influence government decision-making that lies outside the traditional view of "policy" and by adopting a broad interpretation of "policy" it is possible to harness the strengths of a wide array of stakeholders in order to bring influence to bear on multiple decision-makers at different levels within and outside government. Working with the right allies boosts the opportunity for enabling and empowering a wider range of stakeholders, including those whose role is to maintain pressure and influence on both current and future governments.

2

ENABLING COMMUNITIES TO MOBILIZE GOVERNMENT SUPPORT

Sophie Kutegeka and Guyo M. Roba

Introduction and background

In many developing countries, the role of local communities in using knowledge (both indigenous and others) to influence government has been largely ignored. This is caused by a variety of factors including deeply engrained policy positions that eschew the role of communities (Kigenyi *et al.*, 2000). This has started a long-standing public policy debate and academic discourse on the role of community knowledge in mobilizing government and influencing decision-making processes. Consequently, we have witnessed that community knowledge is gradually gaining recognition and is beginning to influence decision-making, particularly in the field of natural resource management.

Historically, Martin Luther King emphasized the importance of people in shaping and influencing the course of events in the society in which they live[1] when he said "I refuse to accept the idea that man is mere flotsam and jetsam in the river of life, unable to influence the unfolding events which surround him." Influencing society involves engaging in dialogue with those around us and with our local institutions, associations and organizations to enrich the quality of our lives and feel empowered to define our destiny. To do so, however, it is argued that people need to be fully aware of what is at stake and what they stand to benefit from their involvement.

Many people in developing countries rely heavily on their immediate environment and natural resources for their livelihoods. However, despite the fact that they are the most exposed to environmental risks and degradation, they are usually the least represented in environmental decision-making. Good natural resource management depends on participatory, transparent, open and accountable governance that ensures the effective participation of the public in the various stages in the decision-making processes (Nyangabyaki, 2003). Effective participation should be interactive, and appreciated as a right, with local people participating in joint analysis, development of action plans and the formation, or strengthening, of local institutions (Pretty, 1995). Effective access to relevant information is a key step in empowering citizens to exercise a degree of control over resources and institutions (this point is taken up in Chapter 3 in this volume).

Empowerment can enhance a community's power to lobby and engage local authorities for changes in policy and practice. Rural communities have a rich knowledge about their natural resources and about how to manage them effectively, and with appropriate support they can become important agents of change (Wyckoff-Baird *et al.*, 2000). When communities are empowered, they can be more capable of mobilizing support both from government, development partners and other actors. This increases their capacity to participate in decision-making processes so that their views are integrated into the government policies and programmes.

In Eastern Africa, decentralization is one of the measures that governments have put in place to bring services nearer to the people, giving them an opportunity to participate in the decision-making processes at the different levels. With the existence of the local government structures, even non-governmental organizations have found avenues to work closely with communities and empower them to influence policies at the lowest levels. This is based on the

FIGURE 2.1 A cultural performance to convey a community message on wildlife conservation to government, Guba Dida, Kenya (IUCN, 2008)

realization that communities are too detached from the central governments and it requires a long process to build their capacity to the point where they are able to influence policies at a national level.

This chapter demonstrates the relevance of community power in mobilizing government support and illustrates how this power can be harnessed. The chapter draws on case studies from Kenya, Tanzania and Uganda to demonstrate what it takes for communities to be real drivers of change, what processes are important to mobilize the community power, the success factors, key challenges and specific recommendations. The case studies provide evidence of how policy and legal recognition of community institutions, knowledge and skills can empower communities and enable them to derive benefits from conservation and thereby develop more sustainable livelihoods.

Understanding community knowledge

Communities[2] may be viewed as systems composed of individual members and sectors that have a variety of distinct characteristics and interrelationships. These sectors are populated by groups of individuals who represent specialized functions, activities, or interests within the system (Thompson and Kinne, 1990). Since the beginning of history, human beings have formed communities that share cultural practices reflecting their collective learning: from a tribe around a cave fire, to a medieval guild, to a group of nurses in a ward, to a street gang, to a community of engineers interested in brake design. Participating in these "communities of practice" is essential to our learning. It is at the very core of what makes us human beings capable of meaningful knowing (Wenger and Snyder, 2000).

Community knowledge can be better understood from the basis of shared identity and practices. Since community members share many practices, communication between them can draw on a background understanding or knowledge that doesn't have to be explicitly stated. It is therefore easier to build a knowledge-sharing system based on community life that stays within the community than one that crosses distinct boundaries. Moreover, community membership is the basis for trust, and effective knowledge sharing depends on trusted information. In this case, the information is not only understandable in context but also trustworthy. However, to understand what constitutes valuable knowledge for the community, it is necessary to observe the way people do their work, which reveals what kind of information they most often share and why they value it (Bobrow and Whalen, 2002).

In a community of practice, members are bound together by their collectively developed understanding of what their community is about and they hold each other accountable to this sense of joint enterprise. In this case, members clearly understand and appreciate the reasons why they are together and have a good understanding of their role and contribution to the community (Wenger, 1998).

The role of community knowledge in environmental policy formulation and practice

> How we manage decisions determines how effectively these decisions are used to manage nature and how fair they are to the people affected. When people participate in nature-related decisions that affect them, they are more likely to support these decisions, and the decisions are more likely to be successfully implemented. When they are left out, it is often a recipe for conflict, inequality and environmental harm.
>
> (World Resources Institute, 2003)

Over the last decade there is growing recognition across the world that citizens should play a role in shaping and informing environmental policy. Surprisingly, the perception and preferences of ordinary citizen, particularly those from marginalized groups, are rarely prominent in the process of environmental policy-making. Indeed, the knowledge of most Africans who still live in rural areas, earning a livelihood in small-scale farming, herding and fishing among other activities is widely acknowledged. Their ability to manage the resource base that permits this should not be underestimated (Veit, 1998).

Experience has shown that rural communities are knowledgeable with regards to managing their environment and natural resources. However, this knowledge often goes unnoticed and does not influence government policies or local interventions. Government policies and programmes have only until recently begun to accept the importance of local knowledge and involving local people in making decisions that affect the environment and natural resources from which they derive their livelihood. This realization has contributed to policies of decentralization in many African governments, and the devolution of rights to use and benefit from natural resources to local governments, the private sector, communities and Civil Society in general (Veit, 1998). The real value of participation is that mobilizing the entire community, rather than engaging people on an individual basis or not engaging them at all, leads to more effective results (Braithwaite *et al.*, 1994). In other words, a crucial element of community engagement is participation by the individuals, community-based organizations, and institutions that will be affected by the effort.

Gaventa and Robinson (1999) argue that citizens can shape policies through four alternative routes: first, through covert and subtle form of resistance; second, through advocacy of community based action groups, trade unions, NGOs and other social movements; third, through political means, through more rigorous political representatives and elections and through more revolutionary and confrontational actions; and fourth, through opportunities created by policy-making institutions in attempt to cultivate culture of participation in the formulation of their policies. Despite the fact that most countries in Eastern Africa have established agencies and laws for environmental management there is concern as to whether the different environmental norms and rules are enforced on the ground. For a country to take pride in promoting environmental governance, citizens need to be able to speak out and protect (or cause the protection of) the environment. This realization has led many countries to establish a constitutional right for ordinary citizens to act in pursuit of a healthy environment (Mugabe and Tumushabe, 1999).

Drawing on local knowledge can bring about more informed decisions that serve local people and ecosystems better. However, an assessment conducted in a number of countries including Uganda, Kenya and Tanzania revealed that governments do not provide as much access to environmental information, or as much opportunity to participate in environmental decision-making, as their citizens would like (World Resources Institute, 2003). Despite the various efforts put in place by governments to promote citizen participation in environmental decision-making, there have been a number of challenges. The case studies presented here indicate a variance between the policies and institutions that have been put in place to support citizen participation and what actually happens on the ground. For example, most of Uganda's environmental laws and policies (including forestry, wetlands and environment policies) were developed through consultative processes but their implementation level is low.

The role of decentralization in promoting community power

Decentralization involves the empowerment of local communities through the devolution of political, administrative, legal, fiscal and other powers from central government down the hierarchy to local levels in a country (UNDP, 2006). The term can also be used to mean a system of government in which power is granted to local authorities or a process by which governance is moved from a centralized to a decentralized system (Prud'homme, 2003: in Okidi and Guloba, 2006).[3] In theory, decentralization means devolving power from central government and empowering local institutions that can better discern how to manage resources and deliver services to meet the needs of local people (Jütting *et al.*, 2004). However, there remain questions over implementation, with many institutions continuing to exclude key players like women, the elderly, people with disabilities and youth.

Decentralization was introduced to contribute to development by empowering local people and institutions at every level of society, including public, private and civic institutions, thereby improving access to basic services. Decentralization embraces all tenets of democracy – empowerment, equity, increased capacity[4] for self-governance, transparency, subsidiary, accountability, transfer of decision-making for management of local resources and justice among others (UNDP, 2006).

Box 2.1 ***Legal and policy framework for decentralization in East Africa***

- In Uganda, the legal backing for decentralization is provided by the Local Government Statute of 1993, the Ugandan Constitution of 1995 and the Local Government Act of 1997;
- Kenya has the Local Government Act, the Local Authority Transfer Fund Act No. 8 1998 and the Local Government Reform Program;
- Tanzania has the Constitution, the Policy Paper on Local Government Reform 1998 and the Local Government Reform Programme 2000;
- Rwanda has the National Decentralization Policy (2001) and the Decentralization Strategic Framework 2007;
- In Burundi, a Ministry in charge of Decentralization and Local Development was created in 2007 and in 2008 the Government adopted a Framework for the National Decentralization Policy and a Three-Year Plan for its implementation.

Decentralization has served as a vehicle for establishing local ownership of resources and the decisions that affect them. This is supposed to be realized through bringing government closer to the people, making it more accessible and knowledgeable about local conditions and more responsive to people's demands. For example, in Uganda and Rwanda decentralization introduced a form of direct participatory decision-making at the lowest level of the local government system (Kauzya, 2007). In Uganda's Local Council System, the lowest level (LC1) is composed of all members of the village who are 18 years and above who elect among themselves a chairperson and executive to lead the process of decision-making and implementation. In Rwanda, the lowest level is the cell where all members of the voting age are members of the council and they elect the executive.

Local authorities are more familiar with local conditions due to their regular interaction with them. They tend to act more in line with local preferences and conditions, and their response to local needs is more expeditious as compared to the previous centralized arrangement where communities would have to rely on their political representatives for the purpose of ensuring that their views are incorporated into the national policies and decision-making processes. With decentralization, local people are more likely to be involved in government planning processes through the structures and systems established at the local levels. Consequently, this promotes ownership and hence sustainability of the processes as will be illustrated by the case studies.

Decentralization has empowered the citizens, heightened their awareness of the different custodians of responsibilities, delivered coordinated services closer

FIGURE 2.2 Fish market in Community Managed Conservation Area, Tanzania (E. Verheij)

to the people, promoted creative local resource mobilization, and increased the responsiveness of public investment to local popular demands (Emorut, 2006). Several windows of communication have been opened between the local population on the one hand and the local and central governments on the other hand (Okidi and Guloba, 2006). This has been made possible through devolved political powers, which enable citizens to elect their own leaders who can demand greater transparency and accountability, and directly influence decision-making processes. The devolution of power has enabled people to participate both in planning, supervising and other decision-making processes, some of which are performed through the various committees formed at the different levels, giving different sections of the community an opportunity for their voices to be heard.

Decentralization is important for mobilizing community power because it can provide opportunities for more people, including the under-represented groups in society, to contribute to making decisions that affect their lives and those of the entire society. When people are given an opportunity to propose solutions to the problems encountered in their everyday life it creates a sense of confidence among the communities and greater appreciation of their responsibility to deal with their own challenges. Certain community challenges require a local approach, and decentralization has enabled communities to think innovatively and use their local knowledge in a way that would be compromised through a more centralized system. This is very crucial in building community confidence and its capacity to monitor and influence initiatives within their localities. However, despite providing a framework that supports under-represented groups to participate using their vote, their voice and their direct action by engaging in specific activities, there has been a growing concern about the need to focus on enhancing the capacity of marginalized groups for effective representation and not just focusing on numerical representation (Kauzya, 2007; Brosio, 2000).

It is argued that decentralization facilitates the mobilization of local resources in support of the development process, and enables value added contributions to the provision of services and development efforts, which increases the total value of services provided, or development achieved, from the limited formal resources available. However, despite improvements in service delivery and accountability, a number of challenges remain. For example, the desire to accommodate everyone's interest has been compromised by the inequalities in society whereby the needs of the rich and influential tend to get more appreciation and consideration in the planning processes. In addition, community members' ability to access information, which is supposed to build their confidence and capacity to engage with the relevant power centres, has been compromised by the general culture of secrecy within government institutions. Due to these shortcomings, Civil Society Groups have an important role to play in supplementing government efforts to mobilize communities and build their capacity and confidence to engage in development planning processes and demand accountability (Kundishora, 2009).

Engaging communities to implement government policy: case studies from Eastern Africa

Community engagement is the process of working collaboratively with and through groups of people affiliated by geographic proximity, special interest, or similar situations to address issues affecting their well-being (CDC, 1997). It is a powerful vehicle for bringing about environmental and behavioural changes that will improve living conditions of the community, mainly through partnerships and coalitions. Over the last decade, boundary changes have occurred within the institution of government with increasing devolution of government business to the region and enhanced opportunities for collaboration between all levels of government and communities (Fawcett *et al.*, 1995).

It is widely acknowledged that involving communities in the design and implementation of development initiatives increases ownership and sustainability. Thompson and Kinne (1990) observed that change is more likely to be successful and permanent when the people it directly affects are involved in initiating and promoting it. This is because when people share a strong sense of community, they are motivated and empowered to change problems they face, and are better able to mediate the negative effects over things that they do not control (Chavis and Wandersman, 1990, p. 73).

A key lesson is that despite the wealth of knowledge in the hands of communities and the need to participate in matters that concern their lives, for them to become real players and agents of change, they need additional resources, knowledge, and skills. One of the ways of ensuring that communities develop these skills and experiences is through ensuring that they participate in the development initiatives that give them opportunities to interact with the rest of the community and external stakeholders. Through these engagements, partnerships are created, which help people to mobilize resources and influence systems, change relationships among partners, and serve as catalysts for changing policies, programs, and practices (Fawcett *et al.*, 1995). However, for a community to grow and reach a level of engaging and mobilizing resources from other actors including governments there is need for strategic partnerships and a continuous process of supporting and nurturing.

Improving people's knowledge, skills and confidence is a key factor in building community's capacity to engage other stakeholders. Communities are endowed with a lot of knowledge and skills of how they have always responded to their local situations. This is mainly the traditional way of dealing with their issues that used to work well before the community setting was interfered with. However, with societal changes and migration of people to new areas, peoples' cultures and ways of doing things have been interfered with. A number of development interventions have also brought in new approaches and ways of doing things that require integration of new knowledge with the traditional knowledge and skills. However, this is a long process that requires continuous collaboration and partnership with a number of actors as explained by the following lessons in case studies.

Case study 1: Securing tenure rights through Community Based Natural Resource Management (CBNRM) in Garba Tula, Northern Kenya

Garba Tula district covers approximately 10,000 km^2 of Northern Kenya and is home to some 40,000 Boran Pastoralists. This district has extraordinary biodiversity and is bordered by the Waso-Ngiro North River, Meru National Park, Bisan Adhi Game Reserve, and the *Chaffa* (Wetlands) where the Waso-Ngiro North terminates. However, the full potential to conserve this biodiversity has not been met and people and their livelihoods instead are threatened by wildlife.

Starting in 2009, members of the Garba Tula community began developing a plan of action for Community Based Natural Resource Management (CBNRM). Through expert-facilitated consultations, the community arrived at a common understanding of CBNRM as "a way to bring local people together to protect, conserve and manage their land, water, animals and plants so that they can use these natural resources to improve their lives, the lives of their children and their grandchildren". This is a strategy "to enable willing community members to play a part in improving the quality of people's lives – economically, culturally and spiritually".

The community identified weak land tenure as one of the key obstacles to success. The County Council legally holds land in Garba Tula in trust, but county councils generally exercise a lot of control over allocation of land and are often poorly accountable to local communities, who in turn are poorly informed of their rights. However, contrary to popular perceptions, Trust Land is not government land and it could allow development of a strong form of tenure if the community understands its rights and the legal mechanisms for asserting those rights (Kenya National Land policy, 2010).

Garba Tula residents are strengthening their rights over land by documenting customary laws and encouraging the County Council to adopt them as by-laws. This will contribute to increased tenure security, which will provide a foundation for developing a range of biodiversity investments that are compatible with the local livelihood mainstay: pastoralism. The community is particularly interested in mapping their resources, including wildlife dispersal routes, with a view to capitalizing on their natural heritage by attracting "biodiversity-investments". Through this approach, communities in Garba Tula are setting up a local trust to manage the process and procedure of engaging local government buy-in and are supported by a number of development, conservation and wildlife agencies as well as central government.

The Garba Tula community members have been able to attract the attention of the local county council because of the participatory processes, which they have undertaken to understand and appreciate their situations, and agree on strategies for improvements. The fact that the community members organized themselves and presented their common position to the county council enabled them to demonstrate the importance of community knowledge in addressing issues that directly affect their lives. This process has enhanced the trust and confidence between the communities and the council to work as partners and support each other to achieve the common goal of securing land rights through CBNRM.

Case study 2: Enhancing forest landscape restoration for improved livelihoods through participatory planning processes in Mt Elgon, Uganda

The indigenous Benet people of Uganda have been confined for years to an area of the Mount Elgon Landscape[5] where they have suffered the effects of severe landscape degradation. This degradation has undermined their livelihood security

FIGURE 2.3 Community forest management, Mount Elgon, Uganda (IUCN Uganda, 2010)

and has reduced crop yields, access to clean water and fuel wood amongst other things. This is a serious concern for the community as it has driven them into greater poverty. It has also impacted negatively on the neighbouring Mount Elgon National Park, which has become the source of most of the goods and services that had been lost from the agricultural landscapes.

IUCN, in partnership with Uganda Wildlife Authority, local government authorities and community-based organizations in the area, facilitated a process to mobilize communities and enable them to harmonize and consolidate their knowledge about their environment. They were supported to identify and articulate what they derive from their landscapes and what it means for each one of them for both their livelihoods and the sustainable management of the forests. The key aspect was to promote a people-centred and holistic approach to restoration and management of this important ecosystem to ensure the emanating goods and services are valued by all stakeholders, especially the local communities who are the main custodians.

A participatory planning process was undertaken to determine what needed to be done and to ensure both improved livelihoods and sustainable forest management. The process of discussion and facilitation was very important for local ownership of the initiative and for generating a common understanding and appreciation of the value of the landscape and landscape restoration. The community were able to come to a common understanding on the barriers that hindered them from implementing their agreed actions.

The key outcome of this participatory process was change of attitude among the Benet. They developed a stronger understanding and appreciation of issues they were facing in the landscape and agreed on the need for collective action to address them if their livelihoods were to be sustained. As a result, community land-care bylaws were developed, contours established and agreements for collaborative resource management negotiated between UWA and the Benet for co-management of forests adjacent to communities. In addition, honey and milk markets, which are the main products from the landscape, were improved to strengthen the local economy through a more diverse set of income streams.

The success of the community initiatives has attracted other partners: both government and nongovernment organizations. At the local government level, the National Agricultural Advisory Services (NAADs) which supports farmers observed the success of the milk and honey marketing and the improvements in local livelihoods and responded by providing funds to procure equipment for product processing and packaging. This led to significant increases in production and in the number of families that benefited: the amount of milk marketed by the community doubled from 100 to 200 litres per day, benefitting more than 50 more families, with each family supplying on average 6 litres per day (US$1.2 per day). The community was able to access NAADS support because of the relevance of the interventions towards improving livelihoods and the fact that the work was conducted in partnership with the relevant government agencies (UWA) and local governments who were in a position to link local initiatives to the available opportunities within government.

Case study 3: Community empowerment to access government support in Eastern Uganda

The Bukhofu Yetana Group (BYG) is a community-based organization formed in 2002 in Manafa district, in Eastern Uganda. It brings together vulnerable groups of people infected and affected by HIV/AIDS. The group focuses on promoting improved livelihoods for people affected with HIV/AIDS, the elderly, orphans and the vulnerable including the very poor families and widows. Since these are the most vulnerable people, and therefore heavily reliant on the environment for survival, BYG demonstrates the best ways of using the environment to improve peoples' lives without degrading it. Members of BYG derive pride from the wealth of knowledge they have in using forest products to produce herbal medicine for treating HIV/AIDS symptoms like cough and skin rashes. A key uniting factor for this group is the fact that they are all affected or infected by HIV/AIDS. Another uniting factor is the wealth of knowledge they share about using forest products to strengthen their immunity and improve their health. It is evident that these factors keep the group strong and united, which enables them to generate more ideas to improve their livelihoods.

Between 2007–2010, BYG engaged in a series of learning events including workshops, tours and demonstrations that provided information on the importance of environment to people infected and affected with HIV/AIDS. The emphasis was on the need to take measures to promote sustainable management of the environment because the communities were taking it for granted that resources like forests would always be available for them to harvest herbal medicine. This learning equipped the group with knowledge and confidence to invest in protecting the environment and the key lesson was to ensure that their experiences and lessons feed into the policy processes starting at the lowest council level. The group realized the need to mobilize and promote sustainable environmental management in the community. This was done through demonstrating the relevance of environment in addressing their specific needs that included harvesting herbal medicine for HIV symptoms, accessing clean water from the catchments, harvesting nutritious vegetables, and fruits that are important for their health.

Through the learning process, members of BYG felt empowered by the realization that they had a wealth of knowledge and experience that could enrich government processes at the local level. The community group targeted district leadership in the natural resources, health and production departments. During a meeting the group presented their strategy, key lessons and experiences. This meeting acted as an ice-breaker both for local government and for the community group. Local government officials were surprised by the wealth of experience that existed within the community and observed that they had been overlooking its importance.

The community group continued engaging with local government and presenting proposals for consideration. One outcome was the local government's allocation of a piece of land and tree seedlings to plant trees in the formerly degraded forest reserve (Khamitsaru forest) in Manafa district. Reforestation was carried out with indigenous and fruit trees, explicitly to enable the community group to benefit from forest resources. The community were also included in the Community Agriculture and Infrastructure Improvement Programme with special consideration of people living with HIV/AIDS and in the government's Farm Income Enhancement and Forest Conservation Project to receive more tree seedlings for riverbank restoration. The District AIDS Committee also took a keen interest in the issues raised by the group and opened dialogue with the group for continuous update on the emerging issues on the ground.

In order to achieve these outcomes the group created subcommittees to promote the application of members' knowledge and expertise in order to raise incomes. This is based on the understanding that various group members have expertise in different areas. Subcommittees are responsible for production of fruit wine, herbal soap, fuel saving stoves, fruit growing, vegetable growing, tree planting and hygiene activities. Interactions within and outside the group have built trust and established norms and relationships that define the interaction and participation of each member.

Case study 4: Accessing government resources through demonstrating impacts on the ground: the case of YOVODEA in Kibaha district, Tanzania

Youth Volunteers Development Association (YOVODEA) is a community based organization in Tanzania that promotes the participation of youth in development. The group received financial and technical support to implement environmental conservation and HIV/AIDS prevention activities in collaboration with the communities of Mailimoja and Ruvu wards in Kibaha district, Tanzania. The environmental conservation and HIV/AIDS interventions were implemented through youth groups in Kibaha district. This partnership aimed at improving the understanding of different community groups of the relationship between the environment and HIV/AIDS, and enhancing the participation of youths in promoting environmental conservation and protecting themselves from HIV/AIDS.

FIGURE 2.4 Village environment management planning, Tanzania (IUCN Tanzania, 2006)

Participatory approaches were used by the group to ensure that community members were involved in the project activities for ownership and sustainability. These involved sensitization meetings at the village level, public rallies, community mapping and door-to-door visits of people living with HIV/AIDS in the selected communities. All the activities highlighted the importance of natural resources to people infected with HIV/AIDS, including providing herbal medicine, improved nutrition, water and sanitation. Through a series of learning tours and field exchanges, interest was raised among both community members and government authorities to engage in environmental actions. During one community meeting, the Ruvu village government offered to support YOVODEA to bring more community members on board and demonstrate how they can address the linkages between HIV/AIDS and environment for the good of both people and the environment.

As a result, the Ruvu village government awarded YOVODEA 50 acres of land to set up a demonstration centre for HIV/AIDS and environment. This demonstration centre was used to promote community participation in activities such as planting of different types of trees (for medicinal, fruit or firewood purposes) and developing natural resource based enterprises. Although the practice in Tanzania is for village governments to offer pieces of land within its mandate to groups or individuals at a cost (or through leasing) the 50 acres were given to YOVODEA free of charge because government leadership appreciated the benefits the project would offer to the communities. In addition, due to the relationship that had been established between the project and the village government authorities, the local government dairy farm agreed to provide manure for free.

Strengthening government policy with local expertise

Although, rural communities have rich and highly relevant local knowledge of their environment and natural resources, unfortunately, until recently, most of it has remained unnoticed and has not been systematically documented. As a result, government policies and programmes have until recently not considered local knowledge and involving locals in making decisions that affect the environment and natural resources on which they depend for their livelihoods. In realization of this, many governments have tried to decentralize resource management responsibilities and to devolve rights to use and benefit from natural resources to local governments, the private sector, communities and Civil Society.

However, there are noticeable changes in some countries with the changing roles of communities in conservation. Devolution of roles and responsibilities to local communities and local government is taking place gradually. There is growing recognition of the importance of traditional institutions and local community structures. Increased community participation in decision-making is a pre-condition for promotion of community-based natural resource management. This is shown in the case of Kenya, where there is general political will to decentralize authority and to strengthen legal recognition of customary governance arrangements.

Communities have a lot of information and skills, but they often lack the capacity and confidence to feed that knowledge into relevant policy processes. For them to be able to push their experiences to policy levels and gain confidence to engage the policy-makers, they need consistent nurturing through information sharing, awareness and exposure. Third parties have engaged many community groups using various strategies but the starting point should always be to recognize and appreciate the wealth of information and skills they have as a group. Being able to get an outsider's view and exposure of what others are doing in the same area is crucial for building the community's confidence and capacity to the level of engaging in dialogue with government.

However, reaching this level requires a more empowering and participatory process to establish long-term patterns of community participation. Participation needs to become a default behaviour amongst staff in government and non-governmental organization and rather than an occasional event. Legal mechanisms need to be explored to protect the roles of communities in natural resource management (both planning and implementation) and to safeguard their benefits from these resources.

Many actors have opportunities to build the capacity of communities and to provide avenues for them to influence decisions affecting the natural resources around them. This capacity starts right from the household level, taking advantage of local government arrangements to enact and implement bylaws and ordinances and get involved in government programmes at the local level. With this kind of capacity, communities become keen to understand how their experiences and views feed into the bigger policy processes. They also take greater interest in discovering the opportunities available within government structures and how they can benefit from them. It needs to become the policy in non-governmental organizations as well as in government to ensure that all community work is deliberately empowering, starting with a concentration on promoting high quality and inclusive forms of participation.

3

MOBILIZING LOCAL EXPERTISE THROUGH NETWORKING AND EMPOWERMENT

Victor Kamagenge

The value of local knowledge

There are many different kinds of knowledge and ways of knowing: "tacit knowledge" that we possess but are not consciously aware of; "implicit" knowledge that we may be aware of, but have not yet articulated; and "explicit knowledge" that has been articulated in written or spoken form (Fazey *et al.*, 2006; Reed *et al.*, 2010). Knowledge includes different "types" of information that may have been derived from a range of sources including personal experience, observations and research results. As such knowledge does not necessarily consist of "universal truths" and may depend on the unique interpretation and reality of each individual (Zermoglio *et al.*, 2005).

Neatly defining local knowledge (or indigenous knowledge or endogenous knowledge) and its scientific validity therefore presents unique challenges. Different forms of knowledge inform and are developed even within "science", and scientific progress is marked by the resolution of disagreements between those who hold different knowledge, informed by different methods and epistemologies (Lane, 2001). One proposal for distinguishing between local and scientific knowledge is to distinguish between the "know-why" of science and the "know how" of local knowledge (Lundvall and Johnson, 1994). It is suggested that science concerns itself with understanding the underlying principles and theories behind observable phenomena, for the sake of future prediction, while local knowledge is primarily tacit, informal, context-dependent and rooted in experience and practice (Ingram, 2008; Lundvall and Johnson, 1994; Reed *et al.*, 2010).

The value of local knowledge is not only instrumental (to achieve a goal), but can also be intrinsic: building on local perceptions and beliefs can be part of an empowering process that builds confidence and capacity at local level. For both

FIGURE 3.1 Knowledge has an inherent value as well as being practically useful, Mandera, Kenya (Tilstone, 2010)

instrumental and intrinsic reasons there has been a steady increase in bringing local knowledge into scientific and policy debates, and in building on local knowledge as an alternative to earlier top-down, "transfer of technology" approaches (Long Martello, 2004).

Knowledge can be relevant at different scales, with local knowledge considered by some to be more geographically constrained and less easily generalizable than scientific knowledge. Knowledge about human-environmental systems should therefore be hierarchical and networked across multiple scales, since choices of scale have political implications by privileging some forms of knowledge over others (Ostrom, 1990; Reynolds *et al.*, 2007). Choices of geographical scale and boundaries will influence decisions over who is a stakeholder, and therefore who frames research questions, and whose knowledge is considered valid (Ostrom, 2005; Reed *et al.*, 2010).

Linking scientific and local knowledge can enable scientists to inform more relevant and effective environmental policy and practice to monitor and tackle environmental challenges (Stringer and Reed, 2007). This may be done through equal knowledge partnerships that construct new knowledge on a combination of tacit, implicit and explicit knowledge from different groups, or it may be done by building awareness to enables the sharing of existing explicit knowledge between different groups. However, there can be significant tensions between these different groups that can impede the frank exchange of knowledge, which may stem from entrenched disagreements, a history of disrespect, or from misunderstanding of underlying context or belief systems (Moyo, 2009).

Scientific knowledge is usually afforded greater legitimacy than local knowledge, and is much more likely to inform the initial problem-setting stage of policy development, largely due to the fact that it is recorded and made explicit (Jordan and Jones, 1997). Scientific knowledge may be trusted by those in power as being objective, dispassionate, controlled, replicable and testable (Briggs, 2005). As a result science is thought to be more influential in policy formulation and local knowledge is often trivialized to the point of being ignored entirely (Mackinson and Nottestad, 1998). Chapters 1 and 7 in this volume partially dispute the assertion that science is necessarily influential in policy processes, and there is a tendency in some quarters to pick and choose from science only that which supports a particular policy agenda. Nevertheless, even in mixed research methodologies, where qualitative research is used to explore and access local knowledge and generate hypotheses, which are then tested using more quantitative methods, scientific knowledge is usually considered to be superior and is used to "validate" local knowledge (Reed *et al.*, 2010).

A number of recent studies question whether it is possible, or even advisable, to distinguish between different types of knowledge. Local and scientific categories are social constructions that cannot be simply classified into separate systems of knowledge at the local scale, where people continuously incorporate new ideas into their management systems. Most producers and users of local knowledge do not distinguish between scientific and local knowledge in everyday practice (Bruckmeier and Tovey, 2009). "If it makes sense and can be used within prevailing socio-economic and physical environments, then it is adopted, replacing previous (and now often redundant) ideas" (Reed *et al.*, 2010).

People use their knowledge to understand and interpret experience and observations and to make predictions and take decisions on different issues. This chapter explores the role and importance of mobilizing local knowledge through the process of networking, and the empowering effect this can have. Recognition of the local knowledge and adaptive skills acquired by people should be seen as a vital resource in the development process and for improving the overall knowledge of social and ecological systems. This chapter illustrates how networking has empowered and enabled communities to influence policies and decisions affecting their welfare.

The chapter defines networks and empowerment as used in various contexts, discusses their importance and also examines key functions of networks and networking as used in communication, motivating creativity and in consensus building. It discusses the elements of empowerment, such as access to information, inclusion and participation, accountability and networking, and presents interventions that have contributed towards these dimensions of community empowerment. The chapter concludes with lessons and recommendations for mobilizing community knowledge through networking for empowerment.

Harnessing local ecological knowledge for government decision-making

Local knowledge has been referred to as the knowledge of any people who have lived in an area for a long period of time, and it is the knowledge used by them to make a living in a particular environment. Terms used in the field of sustainable development to describe this concept include indigenous technical knowledge, traditional environmental knowledge, rural knowledge, local knowledge and a farmer's or pastoralist's knowledge (Langill, 1999). Local knowledge is dynamic and evolves in the local environment and it is specifically adapted to the requirements of the users within the prevailing conditions. Like other types of knowledge systems, local knowledge systems have developed and evolved over time. As a part of overall local knowledge system including social and cultural aspects, local ecological knowledge may derive from diverse sources (Joshi and Shrestha, no date).

In natural resource management existing knowledge systems can be largely distinguished into two main categories: scientific knowledge and local knowledge. Scientific knowledge comprises of knowledge generated through scientific investigations carried out mostly by research institutions through carefully designed investigations. Local knowledge on the other hand is mostly derived from farmers' careful observations of various factors and processes and their logical interpretation (Berkes *et al.*, 2000). The process of deriving local knowledge may be seen as less formal than that of scientific knowledge. Ford and Martinez (2000) call this type of farmers' knowledge about their farm and ecology traditional ecological knowledge, a term used to describe the knowledge held by indigenous cultures about their immediate environments and the cultural practices that build on that knowledge.

In many marginal agro-ecosystems, resource-poor farmers have developed complex, problem-solving, technical knowledge as well as knowledge of conservation of biodiversity and ecosystems (Fujisaka, 1997). Local ecological knowledge consists of knowledge derived from real world observations and farmers' experiences of structures and functions related to farmers' priorities and practice. Farmers tend to think more holistically, with limits imposed on their analysis by what they are able to observe and experience. This creates regularities in local knowledge of natural processes across cultures and regularities in how local knowledge contrasts with scientific understanding. Farmers make decisions and take actions as well as develop new innovations progressively based on their knowledge (Berkes *et al.*, 2000).

Ecological knowledge is not the only form of knowledge that influences local decisions: conditions and constraints due to cultural norms or religious obligations

FIGURE 3.2 Community-driven land rehabilitation trial in Rupa, Uganda (Nangiro, 2005)

may influence farmers' decisions to go against their ecological knowledge. External factors such as market forces, government policies and household financial status may force farmers to opt for ecologically sub-optimal management actions. Farmers, on the other hand, learn from the outcome of their actions or from observing the action so of others, and so enrich or refine their overall knowledge. Similarly, knowledge from other sources, such as radio, television or from peer-learning, often enriches a farmer's knowledge (Berkes *et al.*, 2000). The combined knowledge acquired by a particular community is what is referred to as local knowledge in this chapter.

Proponents of integrating local ecological knowledge into management's knowledge base have offered a three-fold argument. First, they suggest that it can improve the understanding of local ecological and social conditions, producing management decisions and policies that are more responsive to these conditions; second, they offer models of adaptive, sustainable wise use of resources; and third, they quell some of the conflicts and mistrust that arise when local expertise is ignored and discredited (Berkes *et al.*, 2003).

Local people have many generations of experience to draw on, with often long-running cultural memory of processes of environmental change and experiences of what has succeeded or failed in the past (Oba, 2009). Research has demonstrated that local communities have managed the environments in which they have lived for generations, often without significantly damaging local ecologies (Emery, 1996). Furthermore, local technologies have the advantage of relying on locally available skills and materials and are thus often more cost-effective than introducing exotic technologies from outside sources. Utilizing local knowledge in research and management plans therefore gives it legitimacy and credibility in the eyes of local people and the scientists (Langill, 1999).

Networking to mobilize local knowledge

Despite the strength of local knowledge, government decision-making has traditionally ignored it (Kutegeka and Roba, Chapter 2 this volume). Assistance is often needed to mobilize this knowledge, including assistance to the knowledge holders (i.e. communities) and assistance to the knowledge users (for example, local government). In the absence of willingness from government communities can achieve a great deal in mobilizing their knowledge, particularly through greater solidarity and improved communication and networking. This is an important role of Civil Society.

The role of Civil Society in sustainable development has grown dramatically in recent decades, for a multitude of reasons, but particularly to promote democratization and accountability and to strengthen the relevance and delivery of development interventions on the ground. Civil Society should be an agent of "communicative rationality", embodying a process that can bring about change in the public sphere. The ability of Civil Society Organizations to bring about change is tied to their ability to generate and communicate rational knowledge (Habermas, 1984).

Knowledge may empower and it may enable a shift in power relations, but it does not automatically make a person or an organization influential. Civil Society plays a critical role in building social capital by creating connections among actors, which increases the stock of trust and exchange of information and facilitates cooperation in society (Coleman, 1988; Putnam, 1993). This is the tacit understanding behind many of the endeavours to promote networking between local communities and Civil Society Organizations in the development and conservation sphere. Communities, as elements of Civil Society in the broadest sense, can greatly improve their prominence by acting in alliance with others, by sharing information with others, and by exerting collective influence on decision-making in government. This recognition has led to a proliferation of networks, as mechanisms that should promote communication and cooperation.

Perkin and Court (2005), see networks as organizational structures or processes that bring actors together. These structures can be either formal or informal structures that bring together network members (the members could be individuals or organizations). The members should share a common interest on a specific issue or a general set of values. This is a broad definition that can be applied to

FIGURE 3.3 Global gathering of pastoralist communities, Yabello, Ethiopia (IUCN, 2007)

a small informal group as well as large organized institutions. Most of the networks when they are functional can particularly fulfil the following functions:

- Communication – the multiplicity of links within a network allows for actors to communicate better. Hence there is the potential for knowledge to be shared interactively across both horizontal and vertical dimensions.
- Creativity – free and interactive communication amongst diverse range of actors offers a fertile climate for creative action.
- Consensus – networks can make use of their many links among diverse actors to build consensus, often circumventing formal barriers. They allow like-minded actors to identify each other and rally around a common issue.

Networks create opportunities for members to collect, analyze and disseminate knowledge and learning. In most cases networks seem to be cost-effective ways to provide goods and services to a large community. When they are member-driven they can generate ownership of research, practice and policy. They also enable members to develop and project a strong collaborative voice and give members the credibility to allow that voice to be heard in the policy process. Networks thus play a support role through community development and learning among the members, and an agency role by developing and amplifying the voice of the members (Mendizabal, 2008).

Empowering communities to link practice and policy: experiences from the field

The term empowerment has different meanings in different socio-cultural and political contexts, which makes its translation and interpretation difficult: the term is often used but less frequently defined. Empowerment has been described as the process by which the powerless gain greater control over the circumstances of their lives. This includes both control over resources (physical, human, intellectual, financial) and ideology (beliefs, values, attitudes) (Flintan, 2008). The World Bank (2002) has defined empowerment as "the expansion of assets and capabilities of poor people to participate in, negotiate with, influence, control, and hold accountable institutions that affect their lives".

In its broad sense, empowerment could be considered as the expansion of community's freedom of choice and action. Here it means increasing community authority and control over the resources and decisions that affect community welfare. As communities exercise real choice, they gain increased control over their social wellbeing. Decision-making by poor communities can be limited by lack of assets as well as powerlessness to negotiate better terms for themselves with formal and informal institutions (World Bank, 2002).

Institutional strategies to empower communities will vary in most cases due to the fact that communities are not homogeneous. Not only that, but since community social, cultural, political, and economic conditions vary, reform strategies must vary as well. For example, strategies to make school management accountable to parents is different from strategies to have community concerns and needs reflected in the local government annual plans and budgets. Each strategy will vary depending on the political, institutional, cultural and social context of the particular community or community group. Strategies also evolve and change over time in any given context. The challenge, then, is to identify key elements of empowerment that persist consistently across social, institutional and political contexts (Bartle, 2000).

There are many examples of empowerment strategies that have been initiated by communities, governments, Civil Society, private companies and other social groups. Successful efforts to empower communities to increase their freedom of choice and action in different contexts, in many cases share some elements. Even though there is no single model for empowerment, experience shows that certain elements are almost always present when empowerment efforts are successful. The key elements of empowerment that must underlie institutional reform are:

1. Access to information;
2. inclusion and participation;

3. accountability;
4. local organizational capacity;
5. confidence;
6. networking (World Bank, 2002).

These elements are elaborated on below with examples from the field of how they can be strengthened. However, each example that is presented could be presented under more than one element and it is clear that these elements of empowerment are closely inter-twined and mutually supportive.

Access to information

Informed communities are better equipped to take advantage of opportunities, access services, exercise their rights, negotiate effectively, and hold state and non-state actors accountable. Without information that is relevant, timely and presented in a format that can be well understood, it is impossible for communities to take effective action. Information dissemination does not stop with the written word, but also includes group discussions, poetry, storytelling, debates and community theatre – among other culturally appropriate forms – and uses a variety of media including radio and television. Laws about rights to information and freedom of the press, particularly local press in local languages, provide the enabling environment for the emergence of informed community action. Timely access to information in local languages from independent sources at the local level is particularly important (World Bank, 2001).

Relevant information can enable people to access public or private services. To ensure greater flow of information, a Network of Environmental Journalists for Lake Victoria was established between 2000 and 2004, with the aim of promoting accurate reporting and effective exchange of correct environmental information on the Lake Victoria Basin. This network was needed to address the numerous cases of misreporting of environmental issues on Lake Victoria. The network created an opportunity for both scientists and technical experts (those who generate environmental information) to work closely with the journalists (those who disseminate the information), to enhance the accuracy of reporting and effectively disseminate the information. By so doing, the network created awareness among the stakeholders and hence influenced decisions towards the management of Lake Victoria Basin.

The Lake Victoria Network of Environmental Journalists brought together three Environmental Journalist Associations in Eastern Africa: the Health and Environment Media Network (HEMNET) in Kenya; the Journalists Environmental Association of Tanzania (JET); and the Environmental Journalist Association of Uganda (EJAU). Each of these networks acted as a focal point to bring together scientists, technical experts and journalists. The Network organized a series of media events to increase accessibility of correct environmental information as well as conducting series of trainings to improve on communication efficiency and accuracy of environmental reporting. The network improved the information that was being reported and assisted the planners to use correct information to plan future development of the basin.

Inclusion and participation

Promoting community participation in some societies is made challenging by a history of exclusion of or conflict between certain groups. Cultural norms can also impede effective participation, perhaps most obviously through the exclusion of women. Sustaining inclusion and participation therefore usually requires changing the rules and establishing mechanisms for participation. Participation may be direct (every stakeholder participating in a process), representational (by selecting representatives from membership-based groups) or political (through elected representatives) (World Bank, 2002).

When promoting participation in decision-making processes, it is necessary to consider who is included, how they are included and the role they play once included. Participatory decision-making is not always harmonious and priorities may be contested: indeed drawing out and reconciling divergent opinions may be the main intention. It is necessary to manage these disagreements effectively

FIGURE 3.4 Participatory resource mapping, Gedaref, Sudan (Jonathan Davies/IUCN)

to ensure that participation is effective and remains inclusive (Holmes and Scoones, 2000; World Bank, 2002).

In Tanzania, the Tanga Coastal Zone Conservation and Development Programme (TCZCDP), initiated in 1994 and closed in 2007, successfully developed and piloted approaches for adaptive co-management. The programme's objectives were livelihoods enhancement and rural development focusing on marine resources management. The Collaborative Management Area (CMA) strategy was adopted when it was realized that the problems of coastal and marine resource management could not be solved unless poverty was directly addressed in collaboration with the stakeholders.

The collaborative management areas comprised the local fishing grounds shared and used by a group of villages. In Tanzania there were no traditional management systems where fishers were allowed to access all fishing areas. However, CMAs formalized and regulated the usage of shared fishing areas under a management structure that incorporated agreed usage and non-utilization zones. Fishing communities were empowered to participate in the identification of the marine resources around them and in their conservation. The programme worked with coastal communities to identify resources to be co-managed and the communities also participated in preparing regulations on the sustainable management and use of resources.

The participation of fishing communities in the identification and management of the marine resources increased their ownership of the management plans. Fishing communities who initially resisted the management plans were engaged by providing them with the opportunity to participate in sea patrols and to report community members who were violating the agreed actions. The understanding and acceptance of the fishing communities to participate in the CMA activities simplified the process of converting the area into a Marine Protected Area in Tanga, which has been recently gazetted in Tanzania (IUCN, 2007a).

Accountability

Accountability refers to the ability to hold leaders, service providers and government employees answerable for their actions, particularly in the use of public funds. Corruption or abuse of public office for private gain is harmful to communities because they have the least direct access to officials. They also have the fewest opportunities to resort to private services as an alternative. There are three main types of accountability mechanisms: political, administrative and public. Political Accountability, of parties and representatives, can be ensured through elections. Administrative accountability (of government agencies) is sought through internal accountability mechanisms within and between government agencies. Public accountability mechanisms hold local governments accountable to the community, and can reinforce political and administrative accountability mechanisms (World Bank, 2002).

In Kenya, the pastoralist community of Garba Tula district have begun to demand greater accountability with regard to the allocation of their land (see Case Study 1, Chapter 2 this volume). Residents of this district follow a predominantly livestock-based livelihood, which depends on extensive rangeland management and seasonal livestock movements between resource areas. Over the past decades critical resource patches have been lost to them as a result of allocation of land, for example, to create conservation areas or irrigation projects. The land is governed under Kenyan Trust laws, through which decisions over the allocation of land should be carried out by the land board, based on consultation with the communities (for who the land is held "in trust"). In reality the process has traditionally been un-transparent and inequitable. Since 2008 the residents of Garba Tula have embarked on a process of strengthening their land rights by generating bylaws to which they intend to hold the newly formed County Governments accountable.

The process of producing these bylaws, based on customary laws that have traditionally been upheld by informal local institutions, is in itself empowering. Greater tenure security is allowing the community to develop communal resource management plans with a greater certainty that they can implement them. Since they initiated the process of developing bylaws, they are very aware of what the rules state and how they should be applied. The community is now developing the skills and motivation to maintain pressure on the local authorities to ensure the bylaws are sustained and are given continuing legal support.

Local organizational capacity

Local organizational capacity refers to people's ability to work together, organize themselves and mobilize resources to solve problems that threaten their wellbeing. If a community is beyond the reach of formal systems, they usually look within for support and strength to solve their common problems. Organized communities are more likely to have their voices heard and their demands met than communities with little organization. When groups connect with each other across communities and form networks or associations they are more capable of influencing government decision-making and they gain collective bargaining power with their services providers.

Local organizational capacity is key for development effectiveness. This is supported by a study conducted in Tanzania by Narayan (1997), which revealed that higher village-level social capital measured by membership in groups of particular characteristics generated higher household incomes. Community organizations, associations, federations, networks, and social movements are key players in the institutional landscape.

The majority of community case studies presented in this book have achieved progress through strengthening organizational capacity. It was the starting point in all four case studies presented in Chapter 2: the cases of YOVODEA community group in Tanzania, the Bukhofu Yetana Group in Uganda, the Garba Tula community of Kenya and the Benet of Uganda. Similarly it was a critical factor in success of the Collaborative Management Area in the Tanga Coastal Zone Conservation and Development Programme mentioned above.

Confidence

Confidence can be developed at a community as well as at an individual level. It can be thought of as an understanding that the community can achieve its goals, based on their willingness and self-motivation and their collective vision of what is possible. There are many examples of confidence building at community level through programmes of empowerment, which have been popularized in recent years through the greater focus on human rights based approaches to development.

In 2008 IUCN and IGAD, the Intergovernmental Authority on Development, organized a tour of IGAD Members of Parliament to visit the Drylands of Northern Kenya.[1] The aim of the tour was primarily to enlighten the MPs as to the challenges faced by people in the drylands and the opportunities for strengthening livelihoods through conservation-based activities. In preparation for the tour, the hosting communities of Isiolo, Samburu and Laikipia North Districts were supported to develop publications and films to show to the visitors. These materials gave the communities greater confidence in presenting their messages and a greater sense of professionalism.

Eight MPs from four of the IGAD Partner States participated in the tour and they were particularly impressed by the capacity of the communities to represent themselves and articulate their concerns and advice. Representatives from the three communities made a powerful case for greater and more effective investments, and these recommendations were echoed by the visiting parliamentarians during interviews on national television and in print media. The recommendations were published in a joint briefing note and presented to the IGAD Council of Ministers in 2010.

Networking

As mentioned earlier in this chapter, networking is a vital role of Civil Society and enables it to fulfil its role in linking actors, knowledge and decision-makers. Networking can help to build social capital by strengthening the exchange of information and building trust. Networking is critical for alliance building and for generating solidarity – either to bring different actors with different skills together, or simply to enable a group of like-minded actors to gain strengthen in numbers. A stronger and more effective community network can result in a stronger community.

This was observed in Uganda through work in support of the HIV/AIDS practitioners and the development of The Environment and HIV/AIDs Network

(TEAN). Local practitioners working with HIV/AIDS victims in their respective organizations were initially approached to explore the relationship between HIV/AIDS and the environment (see Chapter 2, this volume). These local experts and service providers were operating with negligible external support and were eager to improve their coordination, experience sharing and collective action to influence government, both on environmental and other HIV/AIDS related issues.

They were supported to form a network that enabled them to work more effectively together, to cover more clients within a shorter time and to offer improved services at a reasonable cost. Members shared information and lessons learned during the implementation of their interventions and through this they realized that the costs to conduct their activities were reduced despite their activities increasing. Furthermore, they discovered improvements in their work as a result of learning from each other and sharing challenges as well as important information. The outcome for people living with HIV/AIDS was an improvement in services carried out by the practitioners.

Lessons learnt on networking and empowerment to influence government decision-making

Respect, trust, and social relations

The best technical ideas about community empowerment have to be communicated, owned, and defended within a given community. External agencies that aim to empower communities need to respect the values and social relations of the communities they work with. It takes time, skills, resources, and patience to build understanding and trust among key players. Without this consideration in participatory processes, agreements remain fragile as key actors who have opposed each other in the past have little opportunity to build trust or confidence in each other.

The role of community champions

Almost every case of empowerment or change involves strong innovators and leaders within the community. These processes are typically facilitated by a few members and the wider community joins in later, usually after seeing positive results or after being mobilized by the champions. Many people and communities resist changes, so for an external agency to facilitate a positive change within a community these champions need to have consulted and mobilized. To enable community leaders to engage in dialogue with government it is necessary to work with them to develop relevant arguments. However, in some cases these leaders are already quite familiar with government processes and with key actors in government and are already well placed to initiate dialogue. The key role of an external partner is to equip these spokespeople to make a strong case, perhaps using new evidence, and to ensure that they are consulting their communities effectively.

Elements of empowerment act in synergy

Access to timely and understandable information, inclusion and participation, accountability, networking and investment in local organizational capacity and other elements all reinforce each other to deliver better community empowerment. While much progress has been made on participation, other principles such as access to information, downward accountability mechanisms, and local organizational capacity seem to be at least as important.

Information dissemination is often a missed opportunity

There are many simple opportunities to improve the dissemination of information through community-level project interventions. Information can be shared on useful policies, policy processes, markets, weather forecasts, local government plans, village plans, expenditures and much more. Such information can help the community to influence both accountability and policy dialogue. When communities have regular access to correct information it is easier for the community to hold the leadership or government employees responsible on whatever matter. In addition, access to information can enable a community to engage in a more informed way in policy dialogue and thereby bring their experiences, perceptions and advice into policy-making.

Community capacity is easily underestimated

Community capacity to make rational decisions, effectively manage development resources and to self-empower is often underestimated. Given the opportunity, communities often manage decisions more efficiently than other external agencies. Communities may lack incentives or the funds to take the initiative, although often it may be lack of awareness of a particular process and the opportunities to engage with it. This chapter has highlighted examples of the strength of community capacity to directly influence issues that affect their wellbeing. The Garba Tula community, for example, has gained confidence through its engagement with local government and has thus found the process of self-empowerment to be self-reinforcing.

Improving economic status facilitates empowerment

Access to productive resources alone is not enough to empower a community, although economic empowerment is sometimes considered to be a goal in its own right. The greatest impact has sometimes been achieved when a community's improved access to assets has been complemented by relevant capacity building tailored to the community felt needs, roles, appropriate extension services and information. Such interventions and activities need to be carefully thought out and integrated in order to support the whole empowerment process. Empowerment efforts need to be linked to people's ambitions, lives and livelihoods. Therefore, empowerment has to come from within the community and it has to fit with what people want, need and do. It should not be addressed as a separate movement but be made relevant and inclusive to community systems.

Conclusion

There is clearly a strong case for ensuring that local experience and testimony is reflected in government policy and practice, and a growing number of cases whereby government and other actors are endeavouring to listen to communities. Challenges remain in linking scientific and local knowledge, in empowering communities to mobilize their knowledge and in enabling scientists to both respect and use local knowledge. The habit of engaging routinely with communities and drawing on their knowledge is not well developed, and some government advisors deride local knowledge and question its legitimacy. Government may still prefer to turn to scientists rather than local communities for policy advice, since scientists have the relevant information available in the acceptable format, and are also trained to communicate that in a way that government expects. However, communities can find ways to use their knowledge more effectively, through scientists but also through other actors, including elected representatives who may be more influential in setting overarching policy agendas.

Networking has been demonstrated to be effective in enabling communities to mobilize their knowledge and in empowering communities to take their experiences and advice to the necessary levels in policy dialogue. Empowerment has sometimes been an unplanned outcome of networking initiatives, but increasingly it is an explicit goal. Networking amongst communities has been effective for influencing local government decision-making and is increasingly starting to influence decision-making at the national level. Depending on how communities wield their knowledge, it can influence both setting of policy agendas, formulation of policies and also implementation and evaluation of policies. To do this effectively requires better understanding of which other allies, in and outside of government, communities need to network with.

4

WHO REPRESENTS PRACTITIONERS IN THE POLICY PROCESS?

Ben Wandago

Representation and leadership

The task of policy-makers, in finding out objectively and systematically what works or does not work and developing the appropriate policy, is overwhelming. Local stakeholders have a lot to gain from influencing policy, but may have little or no opportunity to engage with decision-makers, or may lack capacity to exploit the opportunities at their disposal. For these people, their leaders could play a vital role in representing their opinions and interests. This chapter focuses primarily on the role of elected leaders, such as parliamentarians and councillors, as well as traditional leaders, in influencing policy and practice. The chapter considers the different types of leadership that an external agent may work with and presents examples of the strengths and weaknesses of these leaders in influencing different stages of the policy cycle.

Leadership is "a process whereby an individual influences a group of individuals to achieve a common goal" and leaders can play a role in directing an institution or process in a way that makes it more cohesive and coherent (Northouse, 2010). Leaders may be chosen through formal electoral processes (Members of Parliament, Senators or Councillors) or they may inherit their role or gain their role through customary processes (traditional leaders such as chiefs or elders). Leaders may have no formal title but play a leadership role within a community, as in the case of opinion leaders and "innovators" in agricultural communities. These different leaders play different roles, particularly in relation to policy processes, and understanding the role that each plays will enable more effective engagement in policy influencing.

Members of Parliament are elected leaders who are considered formally to be the people's representatives. Parliament as one of the institutions of government

FIGURE 4.1 Visiting Kenyan MP invited to plant a tree in the Guba Dida Community Conservancy (IUCN, 2008)

performs critical functions in governance and policy-making. It acts as an avenue for the citizens to have their voice heard in the management of national affairs through their chosen representatives. It also acts as a check on the Executive branch of Government through its oversight role.

Traditional chiefs are the leaders of a tribal society or chiefdom and they often represent their community through a form of self-governance. Tribal chiefs may include Chairs to a Council of Elders, War Chiefs and prominent spiritual or medical leaders. In East Africa only Uganda still formally has Kingdoms, although these are more ceremonial or cultural than political.

Leadership is often also provided by *interest groups,* which are important actors in most political systems. These groups share a common desire to effect policies that benefit their members or a segment of the society. They are also a means of channelling citizens' concerns to policy-makers. The degree of accountability and legitimacy of interest groups is sometimes called into question, for example, where a group can be influential above the weight of its members due to its greater financial resources or political awareness and connection. The term *Gate Keeper* is sometimes levelled at these groups, meaning a person (or institution) that controls access to something and who is often viewed negatively as one who safeguards particular vested interests. Some creative leaders are effective Gate Keepers as they guard the interest of the communities they represent and thus play an important role in ensuring that only policies that benefit their people are implemented.

Parliamentary structures in East Africa

Parliament is a critical institution in East African governments that provides an avenue for citizens to raise their concerns in relation to management of national affairs. Through citizens' representatives (Members of Parliament) parliament provides oversight on the executive arm of government. The legislature is one of three arms of government: the others being the executive and the judiciary. It performs three closely related roles of *Representation, Law-making,* and *Oversight.* The roots of the East African legislatures can be traced to colonial government systems, and in the case of Kenya, Tanzania and Uganda this is the Westminster system inherited from Britain (Friedrich Ebert Foundation and Legis Consult, 2003).

The parliamentary system in Uganda

The first legislative body in Uganda was set up in 1888, by the then Imperial British East African Company (IBEAC) and later on transformed into a Legislative Council (LEGCO) in 1920. The first Parliament of Uganda was formed at Independence in 1962 and since then it has been providing oversight to policies formulated by the Government. The house business is conducted through three committees namely: Standing, Sessional and Select Committees (Friedrich Ebert Foundation and Legis Consult, 2003).

The parliamentary system in Tanzania

Tanzania established its own first law-making organ the Legislative Council (LEGCO) in 1926, which later changed to the National Assembly prior to Independence in 1961. The Tanzania National Assembly has two types of Committees: Standing and Ad Hoc or Select Committees where most business of the house is conducted.

The parliamentary system in Kenya[1]

Kenya's Parliament dates back to 1906 when the colonial government set up a Legislative Council (LEGCO). The current Parliament has four kinds of committees; Committees of the Whole House; Standing or Select Committees; ad hoc Select Committees; and Departmental Committees with different functions.

The parliamentary system in Rwanda

The Parliament of Rwanda is bicameral with a Chamber of Deputies and a Senate and it was put in place, after a transitional period, through elections held in October

2003. The Senate is a body elected by the population whose main functions and powers include representing the population, passing legislation, scrutinizing and overseeing executive action, approving the appointment of state officials and supervising the application of the principles referred to in the Constitution.[2]

The parliamentary system in Burundi

The constitution of the Kingdom of Burundi was promulgated in 1962 allowing the creation of a Senate, which became fully operational in 1965. The current Burundi parliament consists of the National Assembly and the Senate just like Rwanda's but for ethnic balance, the National Assembly has its composition divided to ensure the Hutus occupy 60 per cent and the Tutsis 40 per cent.[3] The members of the Burundi parliament have the right to propose a bill as well as amend a bill in the assembly and also ask questions on behalf of government.

Local government structures in Eastern Africa countries

Local government refers collectively to administrative authorities over areas that are smaller than a state (country). All five East African states have policies that devolve a degree of decision-making authority to local level, which creates new avenues through which citizens can influence government planning that directly affects their lives.

Uganda's system of local government is called the Local Council System. There are five levels of Local Councils, with LC1 the lowest, responsible for a village or, in the case of towns or cities, a neighbourhood. The LC systems should function by channelling concerns up to the level where action can be taken, and channelling directives back down for implementation local. Each Local Council should have a number of identical positions, including Chairman and Vice-Chairman. All such positions in each level of the LC system (from LC1 up to LD5 or district level) are elected through ballot. However, District Commissioners are appointed by the President of Uganda and the relationship between the Resident District Commissioner and the Chair of the LC5 is not always clearly defined.

Rwanda is administratively divided into Intara (provinces), Akarere (districts), Imirenge (sectors) and Utugari (cells). The district is the basic political-administrative unit of the country, with representatives being democratically elected. It has been created as the centre for the delivery of services for local people. The Akarere Council is the policy-making and legislative body, functioning as the parliament of the district through which people can influence decision-making. The Akarere Executive Committee is the day-to-day contact point between the people and their elected council on issues related to service delivery and development. Decentralization is also being pursued below the District level, under the community action planning or *ubudehe mu kurwanya ubukene* approach.[4]

In Tanzania, decentralization was initiated in as a result of the Arusha Declaration (Nyerere, 1967), which outlined the principles of Ujamaa (Nyerere's vision of socialism) to develop Tanzania's economy. The declaration proposed a villagization program for national renewal and self-reliance. This programme aimed to change rural development by gathering households into administrative units that would be the focus of service delivery. The villagization strategy was designed to promote the role of political leaders and also strengthen participation of citizens. In 1972 the Local Government was abolished and replaced with a direct central government rule but later re-introduced again, when the rural councils and rural authorities were re-established and they became fully functional in 1984.[5] Below the local authorities there are a number of democratic bodies to address local development needs, such as the *Vitongoji,* which is the smallest unit of a village and is composed of an elected chairperson who works with an advisory committee.

Traditional institutions and governance in East Africa

Traditional leadership positions are determined by heritage or appointment and often there are elaborate processes for training, inauguration and initiation, and different mechanisms for accountability. Unlike formal government systems, in local traditional leadership systems, the three kinds of powers (judicial, legislative and executive) may be vested in one person, or may be shared by several people, but are not necessarily differentiated. Traditional systems cannot be assumed to be fully representative, or necessarily to be supported by all members of a community. Neither can they be assumed to perform well in their leadership role. However, they have existed for a long time and clearly play an important role in local governance, despite the fact that many of these governance systems are threatened by rapid modernization and social and political changes (ECA, 2007).

The recognition of traditional institutions in Rwanda is considered as a unique policy of promoting citizens' collective action in partnership with a government that is committed to decentralization (see Box 4.1). The Rwandan approach is designed to increase the capacity of citizens and local government to solve their problems through the principle of participation and collective action. The policy of working with local institutions is intended to strengthen democratic processes and ensuring greater responsiveness to local needs, aspirations and culture. Less formalized approaches to working with local leaders are found in other Eastern African countries, as for example in Kenya and Uganda, where elders and other traditional leaders can play a role in securing communal land rights.

FIGURE 4.2 Local women leaders play a vital role in community-level decision-making: community grain stores managed by women in Kotido, Uganda (Nangiro, 2005)

Box 4.1 *The role of traditional institutions in Rwanda*

Pre-colonial Rwanda was a highly centralized Kingdom presided over by Tutsi kings who hailed from one ruling clan. The King ruled throughout but with the assistance of three chiefs: cattle, land and military chiefs. The chiefs were predominantly, but not exclusively, Batutsi, especially the cattle and military chiefs. While the relationship between the king and the rest of the population was unequal, the relationship between the ordinary Bahutu, Batutsi and Batwa[6] was symbiotic and of mutual benefit mainly through the exchange of their labour.

Ubudehe[7] is the traditional Rwandan practice and culture of working together to solve problems. Under the Participatory Poverty Assessment, each cellule[8] is expected to go through a process of collectively defining and analyzing the nature of poverty in their community. The cellule then goes on to identify and analyze the characteristics of the problems that they face and rank them in terms of priority and action plans developed to address these problems. This framework clearly shows the innovation and the usefulness of the Ubudehe process as a tool that empowers Rwandan people at the grassroots' level to solve their problems. It also fits in perfectly well with the decentralization process thus supporting the implementation of decentralized policies in Rwanda.

Representation in policy process

As discussed in Chapter 1 (Davies, this volume), policy can be defined as a "purposive course of action followed by an actor or set of actors". As such policy can include actions and institutional norms from national to local level, in government as well as non-governmental institutions. The role of different actors in the policy process needs to be carefully scrutinized: the influence of citizens and of their representatives will be very different to the influence of researchers or government technical staff. In particular we would expect different levels of influence over setting of policy (or political) agendas, formulating policy options, drafting policies and developing policy instruments and in the final implementation of policy.

The role of parliamentarians in the policy process

The five parliaments of Kenya, Uganda, Tanzania, Rwanda and Burundi have the triple mandate as avenues for representation, law making and oversight over government activities, which presents a number of opportunities to influence policy and other decisions.

In Tanzania, Parliamentary committees have no power to initiate legislation and their recommendations and resolutions are not binding on any government agency. However, once a Bill is presented before the House, they have the power to summon government officials to either clarify any issue or to censure them in case of any wrongdoing. The Tanzanian case is unique as the policy agenda is set elsewhere but the politicians have a strong influence at the formulation and adoption stages of the process. An example is the recently enacted Mining Act 2010 that was initially led by government technical staff and later taken over by Civil Society with the support of a few champions within parliament. The Tanzanian Parliament was engaged at the committee level where sessions were held internally before the findings were tabled in open parliamentary sessions for debate. Since 2010 was an election year there was political pressure to pass the bill given the importance of mining sector in Tanzanian economy, but Civil Society Organizations worked effectively through parliamentarians to persuade government to include their concerns (Olan'g, 2010).

In Kenya most legislation originates in Bills introduced after Cabinet approval. Consultation takes place while legislation is being formulated, and ministers and civil servants often consult a variety of experts, interest groups, trade associations and others likely to be affected by the legislation. Out of the 25 major legislation passed by the Ninth Parliament (2003–2007), 21 were from government while only four were from Private Members' Bills (Barkan and Matangi, 2009).

In Rwanda the introduction of bills is the prerogative of the Government and Members of the Chamber of Deputies, which sets the agenda for policy formulation. Ordinary MPs can propose amendments to these bills at the committee stage before the final decision is taken at the plenary. An example is the process followed in the passing of the gender-based violence (GBV) in 2006. The bill was prepared by Rwandan Women Parliamentarians through national consultations with different stakeholders with support from UNIFEM and UNDP. The bill was adopted in the plenary sessions and is now law that protects women from GVB that was common after the 2004 Genocide (Zirimwabagabo, 2007).

Overall it appears that parliamentarians have the greatest opportunity to influence policy at the agenda-setting stage, although this is often dominated by the ruling party, and at the point of formulation of the policy or resultant legal instruments, particularly through participation in parliamentary committees. Parliamentarians may have limited influence over policy implementation, unless through influence brought to bear at a local level.

The role of Local Councils

Local governance structure in the five East African States allows for elected leaders to participate in the policy process at a local level. Councils are semi-autonomous corporate bodies that to some extent set their own agendas in terms of policies and ensure that implementation is carried out according to the resources available. For example, in Uganda it is the responsibility of the executive committee to initiate and formulate policies for approval by the Local Council and then monitor and oversee the implementation of policies and programmes in addition to making recommendations to the council persons to be appointed members of statutory commissions, boards and committees (Mugabi, 2004).

The devolution of powers through councils is intended to improve service delivery by shifting responsibility for policy implementation to citizens, promoting good governance by placing emphasis on transparency and accountability in public sector management. Devolution of powers is also intended to develop political and administrative competence in the management of public affairs. In Rwanda, the *Akarere* (District) Council allows elected leaders to fully participate in decision-making at the District level and eventually influence the National Policy process by setting the agendas. In Kenya and Tanzania the District councils enjoy significant autonomy including generation and use of revenue that allows elected leaders to set their development agendas and oversee their implementation.

The role of traditional leaders

The mandate, or anticipated role, of traditional leaders cuts across all the steps in the policy process. Traditional leaders are often selected through rules of succession and have a big say in who gets the vote at grass root level, so politicians often ensure their cooperation to promote new policies for implementation. In Uganda the government's own technical advisors have a strong influence in agenda setting during the policy development process and politicians are able to give input at

FIGURE 4.3 Training communities in participatory video to influence policy in Pangani, Tanzania (credit Katharine Cross/IUCN)

the committee stage. Lobbying at this stage is very critical before the policy document is passed by Parliament. It is at this stage that Traditional Leaders can exert significant influence through their MPs, as demonstrated by the Institution of Traditional and Cultural Leaders Bill 2010 that was introduced through the Minister of Culture but received a lot of resistance from Cultural leaders and resulted in an amendment that saw 15 out of 21 clauses revised or deleted (Ultimate Media, 2010).

Traditional leaders may exert influence over policy, and broader political, processes through their prominence in society. For example, they may be very influential over the selection of local parliamentarians and can use this influence to dictate the actions of these parliamentarians. Local leaders are also often nominated to represent a given community in government and non-governmental consultations and through this respect are afforded considerable influence. The precise nature of this influence differs greatly according to both the society and the nature of political processes in different countries and further insight is needed to understand how genuinely influential local leaders are in this respect.

Examples of engagement with leaders to influence government policy and planning

MPs tour of Pangani River Basin

In September 2007, IUCN organized a tour for East African Parliamentarians to the Pangani River Basin in Tanzania. The tour was attended by 24 participants drawn from the East African Legislative Assembly in Arusha, and the National Assemblies of Kenya, Uganda and Tanzania, all of whom are members of the Committee on Agriculture, Environment and/or Natural Resources. The tour was designed to raise awareness of the status of water resources in the region and efforts that are underway to ensure a sustainable water future. The trip consisted of field visits to three sites (Nyumba ya Mungu Reservoir, Lake Jipe and a small irrigation scheme in Kivulini) and a series of related presentations. The participants also had a chance to watch the renowned film on climate change entitled *An Inconvenient Truth*.[9]

Some of the key issues that arose from this tour include:

- The infestation of Lake Jipe by invasive species including *Typha domingensis* and *Cyperus papyrus* has caused siltation and has harmed the livelihoods of the local people as fishing grounds have become inaccessible. As a long-term measure, the MPs recommended the removal of the weed. In the short-term, they proposed to explore options for communities to learn to utilize the weed for income generation, e.g. in making mats, furniture, etc. Since the lake is shared between Tanzania and Kenya, the MPs noted that the implementation of the proposed joint initiative (which has been developed under the coordination of Pangani Basin Water Office and Coastal Development Authority) should be fast-tracked to address these problems.
- At Nyumba ya Mungu, MPs expressed their concerns about on-going construction by a private developer without an Environmental Impact Assessment (EIA) being carried out as per the law. They also noted with concern the low standards of living among communities residing around the dam and the fact they had no access to sanitation facilities, potable water and electricity despite the fact that they live by a dam that generates electricity.
- At Kivulini Small Scale Rice Irrigation Scheme, the farmers explained that they wanted free water for irrigation as they were protecting important catchment areas. The Pangani Basin Water Office explained the need to amend the laws to cater for payment for environment services (PES) so that the community can receive the water for free. Many questions were raised around the modalities of payments and amounts to pay. The Parliamentarians were requested to propose amendments to the water and other relevant Acts in support of this proposal.

At the end of the trip the MPs agreed to bring to their respective Parliaments motions related to water and environmental issues for discussions and formulation of laws. This was in response to on-going policy reforms in the water sector being undertaken by the Tanzanian Ministry of Water and the need for policy harmonization to manage shared resources like Lake Jipe. The MPs requested further assistance for members of the Parliamentary Committee on Agriculture, Environment and Natural Resources from the five East Africa Legislative Assemblies plus EALA (in Arusha) to have a joint session to develop strategies on how to influence policies within their mandates. As with all the case studies presented here, a great challenge was faced in following up on the actions of MPs and tracing their influence on subsequent parliamentary discussions.

IGAD Parliamentarians tour of northern Kenya

Dryland communities often display a high degree of respect towards their environment and, given the choice, protect plants and wildlife even when they do not derive direct material benefit from them. Such services to the environment continue being lost however, particularly when the communities who provide the services are excluded from enjoying them. Enabling drylands communities to invest in and derive benefits from these 'externalities' of their production systems is crucial for both reinforcing their livelihoods and ensuring that the environment continues to be protected.

In October 2008, the International Union for Conservation of Nature (IUCN) and the Intergovernmental Authority on Development (IGAD) facilitated a tour of parliamentarians (MPs) and journalists from IGAD partner states to the drylands of Kenya, to learn first-hand from local communities about the challenges and successes in conservation and sustainable livelihood development. The tour was organized in partnership with the Kenyan Arid Lands Resource Management Project (ALRMP) and the Centre for Minority Rights and Development (CEMIRIDE) with the aim of bringing parliamentarians to meet communities who report success, in their own terms, in sustainable development and using conservation as a livelihood strategy.

During the tour the MPs met members of these communities to observe the outcomes of success, to learn from the communities how they define success, and to draw out lessons that could either support further good practice and innovation, or could guide policies that support these developments. At the end of the tour, the parliamentarians were given the opportunity to examine their own lessons learned during the week, and to propose ways that these lessons could be translated into appropriate action, within their own countries and in the wider IGAD region. One of the MPs made a statement in support of local land use strategies and calling on the Government of Kenya to amend the Wildlife Act to include compensation for livestock lost through predation and also improve on water availability, security and infrastructure (roads) in the conservancies as part of diversifying income from tourism.[10]

These discussions led to a number of recommendations that were later transformed into a policy briefing paper for presentation to other Parliamentarians and Ministers within the region, through the IGAD Council of Ministers. The recommendations were also picked up in national print and broadcast media. Key recommendations from the tour included:

- The participating MPs to share their experiences in their home countries with a view to influencing their colleagues to adopt best practices through policy guidelines, especially on Community Based Natural Resource Management.

- Government should decentralize more responsibilities and decision-making to the local level and encourage more local conservation initiatives using indigenous knowledge.
- IGAD Partner States should promote sub-regional policies for managing shared (trans-boundary) dryland natural resources as part of policy harmonization through the bilateral committees that are already in place.
- IGAD should create mechanisms for information flow and exchange especially through the media.

During the tour, there was keen interest from the members of parliament on the importance of supporting communities in managing their resources especially through conservancies, which if well managed would be a major source of revenue. One of the immediate actions that MPs pledged to undertake after the trip was to report to their parliamentary committees on Environment or Natural Resources or similar on their findings with aim of influencing policies that govern resource allocation to ensure that drylands are allocated more resources.

Recommendations from this tour were summarized in a policy briefing note that was shared among the MPs and government advisors through a conference of IGAD Directors of Economic Planning and Conservation. The combined recommendations of the parliamentarians and technical advisors have influenced IGAD debates at the Council of Ministers and have been adopted in recent IGAD initiatives on Trans-boundary Biodiversity Management.

Further opportunities for the IGAD Secretariat to present the recommendations to the Council of Ministers[11] for adoption and ratification were overshadowed by insecurity in one of the Member States (Somalia) and throughout most of 2009 and 2010 IGAD held special sessions to address the Somali peace and reconciliation efforts. This highlights the challenge of working with parliamentarians whose priorities shift quickly according to changing issues in their countries and region.

MPs tour to Mt Elgon

Mount Elgon, which straddles the border of Kenya and Uganda, was identified by the Eastern African Community (EAC) and Partner States in their 2001–2005 EAC Strategic Plan as a trans-boundary ecosystem that needed to be managed through a regional programme of conservation and sustainable development. Accordingly, the Mount Elgon Regional Ecosystem Conservation Programme (MERECP) was a response to the need for a regional approach to the management of this trans-boundary ecosystem as an important water catchment for Lake Victoria, the River Nile and Lake Turkana. The programme falls within the framework of the Lake Victoria Basin Commission (LVBC)[12] Operational Strategy 2007–2010 under the Environment and Natural Resources programme area.

The East African Legislative Assembly (EALA) Public Accounts Committee (PAC) visited the Ugandan side of Mt Elgon from 9 to 12 March 2010. The purpose of the visit was to make an on-the-spot assessment of the Mt Elgon Regional Ecosystem Conservation Programme (MERECP). The tour was conducted to improve understanding of the functioning of MERECP as a Programme and the challenges in its implementation.

The tour was informed by the report of the PAC on the EAC Audited Accounts for the year 30 June 2008, and specifically by the item on audited financial statements of MERECP. The EALA Public Accounts Committee recommended to the Assembly to visit the Districts of Sironko (Uganda) and Mt Elgon (Kenya) plus other MERECP Project sites to inspect and confirm the ongoing implementation.

Some of the key issues that emerged from the on-spot assessment by the EALA PAC include:

- Livelihoods interventions in MERCEP are of value but need better coordination and integration.
- Monitoring of interventions needs to be improved and sustainability needs to be ensured through greater support and input from the Districts.
- Cross-border issues for Kenya and Uganda, such as harmonizing policies and legislation for joint management of the Mount Elgon ecosystem, need to be clarified and given support from the EALA.
- Greater engagement is needed with the EALA Committee on Agriculture, Environment and Natural Resources, which would spearhead the harmonization process with support from government technical staff.

The programme has since been redesigned but there remain challenges in taking the recommendations forward. Feedback of the recommendations into regional and national parliaments was not well performed, and was difficult to monitor, while the cross-border concerns have not been pursued by the EALA committee on Agriculture, Environment and Natural Resources. It is clear that working with Parliamentarians needs to be carefully examined to look for improvements that can ensure the enthusiasm shown on tours translates into follow up action.

Lembus Council of Elders[13]

Lembus is a common word adopted in the Tugen language, from the corrupted English word meaning "*Land Bush.*" The Lembus Forests are one of the few remaining indigenous forests in Kenya, comprising of gazetted forest blocks (16,211 ha) and unreserved but smaller forest patches of about 120 ha in the Northern Mau complex. In terms of local value, Lembus forests are of immense water and soil conservation value, acting as one of the water towers with a watershed for river systems of the Rift Valley and Western Kenya. However, there is a very high level of degradation in these forests, mainly as a result of human activity, characterized by a history of exclusive discretionary concessions at the expense of indigenous rights and good forest conservation practice in and around the forests.

In response to these threats, and to increase local community involvement in the conservation efforts, the Lembus Forests Integrated Conservation and Development Project (LFICDP) was created. The Lembus Council of Elders (LCE) is one of the key community interest groups in this initiative. The LCE members are a repository of information that is valuable for the conservation and management of the Lembus Forest and are key to the involvement of local community. Through the LFICDP elders hope to pass on their experience and knowledge for the future security of their forest. In the words of one of the LCE Executive members Mr. Philip Sura:

> When we were young boys and warriors, our fathers and forefathers taught us about our culture, the land, the trees and all animals in the forest. Through the most respected elders, they also talked to the white man (Mzungu) when he arrived about our relationship with these forests you see here . . . and he (white man) understood how much we valued the forests and everything therein.

When Mzungu went back to his country, one of our respected elders was given the 'hat' as local administrator, Chief. He and his successors related well with us and passed our culture and values of the forest to our children through local meetings (Barazas). However, in the recent times, the local administrator is no longer chosen from the respected elders . . . the 'hat' is taken up by anybody irrespective of his status in our community!

Our values, history and culture, as we were given by our forefathers, have not reached our children and grandchildren . . . we can no longer entrust these on the 'wearer of the hat' alone. It is now time that we want to directly pass all our undiluted culture to the young generation, because the environment is crying to us that we did not teach our children about the relationship between the Lembus forests and their survival!

By October 2008, the LCE had emerged as a key player in Participatory Forest Management (PFM) of Lembus. The institution strongly influences implementation of Forest Policy at the local level by encouraging community participation in forest co-management. The Elders have continued to gain recognition from the Kenya Forest Service and Kenya Wildlife Service and from the County Council, which is becoming important in policy processes with the pending devolution of power over management of natural resources.

Politicians have also not been left behind and have been using this Institution during their peace and reconciliation rallies in the area. The areas surrounding Lembus Forest in Koibatek District was greatly affected by post-election violence in 2008 and some project staff had to be relocated. The Elders were instrumental in restoring peace after the chaos and enabled the Lembus Project to complete two water installations belonging to two different ethnic groups that were caught up in the conflict.

Strengthening representation in policy processes

Parliamentary structures in East Africa provide numerous opportunities to influence government policy and practice. Members of Parliament have formal opportunities to influence policy formulation and the development of policy instruments and more informal opportunities to influence their implementation. MPs can also influence practice and policy prior to their development, by setting policy agendas and championing causes. An example is the debate over planting of eucalyptus trees in Kenya where the Minster of Environment declared that the tree should be uprooted along the rivers and the Provincial Administration started implementing the directive without any guidelines until Kenya Forest Service. The Forest Service was then quickly forced to develop Eucalyptus Guidelines to advise farmers on where to plant which variety.[14]

External agents may try to influence government policy and planning by working with parliamentarians, but the examples presented here illustrate some of the challenges that can be faced. One of the main challenges has been tracing influence on MPs through to their influence on policy and practice. MPs are lobbied by many actors over many issues and it is challenging to make a particular issue the priority for them. The IGAD MPs tour presented above highlighted the influence that communities can have over MPs if they are supported to make their case effectively and it may be this partnership between external agents, community members and their elected leaders that is most effective.

Given the transient nature of many elected leaders, it may also be most effective to engage them on clearly defined issues that will deliver short-term goals, or at least actions that are visible in the short-term. General training and exposure of MPs to lessons from within their country or region may help influence parliamentary debates, but to ensure that this is effective the training and exposure need to be linked to real opportunities within parliament. This degree of responsiveness and opportunism is not easy within many project-based interventions. On the other hand, engaging MPs in issues within their own constituency, as in the Lembus Case Study, may be effective in influencing local practice and also raising local concerns in parliament.

At a local level, authorities can influence local government agendas and investments. Acting within the powers delegated by central government, they can formulate bylaws and other agreements or plans to guide the use and management of natural resources and the wider environment. Current structures allow local authorities to set their own development agendas within certain legal limits and to implement programmes based on available resources.

Engaging local councils has been challenging in the initial stages of project development, since implementation plans can be derailed by the efforts of local leaders to redirect activities. It takes a considerable level of dialogue to involve local leaders in a constructive role and to develop a working relationship. However, by involving leaders earlier in project design some of the challenges can be reduced. Building working relationships with local leaders through programme implementation in their areas makes it easier to influence their perceptions and priorities, which can lead to long-term influence over local policy-making processes.

Participation of local government leaders in planning and implementing initiatives with communities on the ground, although painstaking, can be considered a legitimate form of communication (Chapter 8, this volume). It can be highly effective in changing the way these leaders think and act. Institutions of Local Government vary greatly between the countries but the autonomy they enjoy, plus the leadership structure that permeates to the lowest level-grassroots, makes them potentially very effective in influencing both policy and practice but mechanisms have to be put in place to exploit this opportunity.

The principle of communication-through-participation, which could be considered as a form of action learning, applies equally to work with traditional leaders. A number of initiatives have worked with traditional leaders to help them influence government policy and planning, but the outcomes have been mixed. However traditional leaders can influence government leaders in a number of ways and can be useful champions of local issues and local good-practices.

Traditional leaders have also been shown to be influential at higher levels of government through their relationship with elected representatives. Traditional leadership can play an important role in election of politicians in the region and Councils of Elders in many countries have reinvented themselves and are now playing a key political role. Although there may be discomfort in some parts of government over this role, traditional leaders are just one of a number of local lobbyists that may seek to influence policy through parliamentarians and this role should be recognized and understood.

Traditional leaders play a vital role in their community and there is scope to develop good practices from their culture and use this to influence policy. The involvement of traditional leaders in policy processes can also be empowering and can ensure that community views are heard and considered. In Eastern Africa there is great diversity in the nature, motivation and legitimacy of traditional leadership, and in many cases they can represent local and indigenous knowledge and institutions more effectively than other policy actors. However, it is often necessary to build capacity for more effective representation (for example, of women) and more effective engagement (for example, through better understanding of policy processes).

Policy debates often take place at short notice, which often confounds the intentions of Civil Society Organizations to engage in them since they lack the time to mobilize the necessary resources. Traditional leaders however can be well placed to act at short notice, and they can be prepared for this through ongoing support and capacity building from Civil Society Organizations. Traditional Leaders can therefore be prepared to influence government at any given moment, and also to sustain pressure through political cycles and changing political leadership. These are roles that local leaders are often expected to play and which project-funded Civil Society Organizations (CSOs) are less well-equipped for.

External agencies work with leaders for a variety of reasons: for greater legitimacy, for improved relations with the community, and because local leaders may be considered to be better connected and more aware of local challenges and opportunities. What appears to be weak in East Africa is an understanding or capacity amongst non-governmental actors to effectively engage with their leaders. As demonstrated in the case studies, more incisive and strategic support is needed to work through these mechanisms to influence policies at national level.

In general, if MPs are influential at the stage of policy formulation, it is vital to influence them at an early stage in a policy process, or preferably before a policy-process is formally launched. Although some policies may be developed over the course of years, with steady input from consultations, others may be more rapid and this creates a challenge for external actors to mobilize the resources and evidence in time.

Organizations can overcome this through steady and ongoing engagement with elected leaders, for example, organizing exchange visits and learning tours, or seminars and other forms of learning, with the aim of changing perceptions and influencing the leaders' thinking for long-term change. There are a number of challenges to this approach however: it is costly to maintain continuous engagement, and often very difficult to demonstrate the results in the time-frame expected by a donor; it is also necessary to sustain such awareness raising through political cycles, which can lead to regular change in elected leadership. Perhaps the biggest challenge observed through the case studies in this report was in following up beyond learning events to track where behaviours and attitudes were changing. Parliamentarians are usually busy, very independent-minded, and not easy to pin down on the sort of monitoring questions that would help learn from such learning events.

Policy influence can be more effective – both in terms of the practical relevance and in terms of local buy-in – when it is conducted in close partnership with communities and their leaders. It is important to focus on inclusion and empowerment of leaders and their constituencies and on convening opportunities for dialogue between these two groups and with different levels of government. It may be necessary to influence government to give greater recognition to the role of leaders, and support is needed to encourage more, and more effective, participation.

A priority for external agents as well as local Civil Society Organizations should be to build capacities at the level of local citizens, whose role is to elect accountable leaders and to influence the leaders they elect. Part of the capacity building and empowerment process is to convene opportunities for dialogue between citizens and their government leaders – both elected and unelected. The principle of multi-stakeholder dialogue should therefore be pursued as a way to ensure that mechanisms are in place to respond more quickly and effectively to arising opportunities to influence policy and practice from local to national levels.

5

PARTNERING WITH GOVERNMENT TECHNICAL EXPERTS TO INFLUENCE POLICY AND PRACTICE

Claire J. Ogali

Introduction

Amongst the many non-governmental agencies that set about influencing government policy, perhaps the majority do so by publishing a study or a briefing paper and passing it to a government employee in the hope that they will act on the published recommendations. It is assumed, sometimes correctly, that these government staff members have the capacity to influence policy. It is also assumed that they will be receptive to the advice and that they can understand and make use of it. In reality, this approach to influencing policy can be haphazard and is often ineffective. For this approach to be successful, it is necessary to understand the role that government experts play in the policy process, including their strengths and their limitations, and to also understand how to build effective working relationships with them.

Government experts are employed by the Executive branch of government, which is also sometimes simply referred to as "the government". It is the branch of government that is responsible for the day-to-day management of the State and includes the legislators, administrators and arbitrators who run the State bureaucracy. The Executive plays a central role in the development and implementation of government policy at all levels, but is answerable to the legislative branch of government (e.g. parliament), as discussed in Chapter 4. The Principle of Separation of Powers is intended to ensure that the executive is held accountable to the people. In theory it is the role of the *legislature* to make laws, the role of the *judiciary* to interpret them, while the role of the *executive* is to enforce these laws (Kettl, 2009; Maravall and Przeworski, 2003).

The Executive is staffed by Public Servants, although the definition of Public or Civil Servant differs country to country. In some cases it includes employees

FIGURE 5.1 Farmer and government extension workers evaluating sorghum varieties in Kotido, Uganda (Nangiro, 2004)

of publicly-owned companies or parastatals. In some countries the term includes local as well as national government employees. It is these government staff that this chapter primarily focuses on, although some sections also discuss government advisors that are not official public servants. The definition of a Public Servant may be quite clear in different countries, but their role in providing technical guidance for defining and implementing policies is not always so obvious.

Public sector technical advisors are individuals with specialized training, who approach societal problems from the vantage point of their knowledge and experience (UNIRSD, 2004). These experts do not work in isolation and often work with, or delegate responsibilities to, scientists, academics and other professionals outside the public sector. Chapter 7 discusses the role of non-government scientists in greater depth. Government experts are expected to provide policy-makers with accurate and complete information backed up by legitimate research, with a view to enabling their leaders to make informed decisions (Johnson, 1996).

Government experts can be instrumental at various levels in the process of formulating and implementing policies. Their involvement in the policy processes is often highly valued since they are viewed as experts in their different disciplines and are expected to be non-partisan and independent of political influence when making their contribution to the policy processes (Khoo, 2010). This chapter examines the role that these experts play in policy and other government decision-making, and examines efforts that have been made to influence this role and to enable government experts to wield more effective influence themselves.

Policy-making roles of some key experts in government

Permanent secretaries

Permanent secretaries are the most senior Civil Servants in their respective ministries or departments and are responsible for the day-to-day running of their ministry or department. They are usually the principle advisors of their Minister and are mostly expected to be highly qualified and experienced in the subjects that their Ministry deals with. Their main functions include organization and operation of the ministry, oversight of the use of public funds, giving technical and administrative advice to the minister and implementation of government policies (Republic of Kenya, 2005).

In East Africa, Permanent Secretaries are usually appointed by the President (see for example, Chapter II, Part 2 of the Kenyan constitution and Article 174 of the Ugandan constitution). They are a critical link between Ministers and the operation of the Ministry and therefore they are instrumental in both influencing and implementing the policies and strategies of the Minister. Since they are key advisors to the ministers they play a key role in "selling" new policy ideas to the ministers and explaining the importance of implementing a particular policy.

Lead government agencies

Government agencies are formed by Acts of Government with the aim of addressing particular areas of public concern and developing and implementing policies within their sectors. Examples in East Africa include the Kenya Wildlife Society (KWS), Uganda Wildlife Authority (UWA), the National Environment Management Authorities of Kenya, Uganda and Tanzania and the National Forest Authority (Uganda). The heads of these agencies in Kenya, Uganda and Tanzania are usually appointed by the Minister, and this process can be politically driven, although it should follow certain guidelines, such as the State Corporations Act Guidelines in Kenya. Where selection does not follow the guidelines, the performance of these Departments, in terms of policy implementation, can be compromised, since the heads may act in the interest of the minister, rather than the public and may not work objectively (Waller *et al.*, 2008).

These government agencies are instrumental in ensuring that policies passed by parliament are implemented from the national to the local level. They are also instrumental in advocating for policies to their respective ministers that would then be forwarded to national assemblies for debate or to parliamentary committees. Some agencies are designed explicitly to have a direct impact on policy and policy processes. The Kenya Institute for Public Policy Research and Analysis (KIPPRA), for example, is a research and capacity building institution that was created to provide quality public policy advice to the government of Kenya and other stakeholders.

Government technical advisors at local levels

Civil Servants working at the local level of government play an important role in the local implementation of policy. Government representatives at the local levels include ministry employees, such as agricultural extension workers, health workers, environmental officers and veterinary officers. These Civil Servants may play a role in reporting back to government and can play a role in policy evaluation. They can also be instrumental in persuading communities in their area to accept certain policy ideas (Khoo, 2010).

Local Civil Servants work closely with local communities and therefore they can play a role in harnessing community ideas to develop local policies and plans that serve local priorities. These ideas may be used as feedback to national governments and incorporated in policy debates at the national level, although mechanisms for this communication are often weak. Working with non-governmental partners however, local Civil Servants can be assisted to communicate local experiences to central governments, and can also add legitimacy to the messages of those non-governmental organizations.

Partnering with government experts to influence policy and practice

Government technical advisors can reinforce the decision-making capacity of government by conveying policy-related information from a variety of sources to the ears and eyes of policy-makers. Their work has the potential to highlight societal problems and provide alternative solutions for them. They can also inspire public debate and add their research and analysis to the voice of the people hence advancing their influence on the policies that shape their lives (Johnson, 1996).

The influence of government technical advisors on the policy process can either be direct or indirect, and can in theory be exerted at any step in the policy process outlined in Chapter 1. Government technical advisors can influence the policy agenda, policy formulation, policy implementation and policy evaluation (Barkenbus, 1998). However, it is evident that influence has to be given to the right experts to ensure an effective result. For example, to influence the policy agenda at its outset it is critical to convey advice to the level of Directors and Permanent Secretaries. To develop good practices in the implementation of policy it is important to work closely with local government experts.

Influencing the policy agenda

Government experts, through their research and advisory work, can influence agenda setting in the policy formulation process by influencing the thinking of senior decision-makers on technical matters. Depending on the level of the expert, they may be able to influence thinking and priority setting of Permanent Secretaries and Directors, or they may be drawn upon to legitimize attention to particular issues, and they can therefore be seen as drivers for issues reaching the public agenda (Dearing and Rodgers, 1996).

In order to be able to influence the policy agenda, technical advisors must influence the general climate of thinking about a policy, and consequently, change the frame of reference of their leaders. This can occur on a number of levels and occasionally the whole framework is altered. An example is the change of thinking regarding environmental issues (particularly climate change) in recent years and the resultant development of policy agendas to address environmental concerns as a matter of priority. The main drivers of these changes have been a combination of scientific investigation and pressure group activity (Mathews,

Box 5.1 *Influencing Government Directors through Experts Conferences*

In an attempt to raise awareness at the highest levels of government on the importance of biodiversity to economic development, IUCN in partnership with the Intergovernmental Authority on Development organized the 2007 and 2010 Conferences of IGAD Directors of Economic Planning and Conservation. The first conference was organized to present a number of national studies outlining the economic value of nature, with particular focus on the neglected drylands. The conference was used to develop recommendations for increased public investment in and attention towards these areas.

The recommendations were adopted by the conference organizers as well as the participants, with an agreement to report back after two years. At the second conference in 2010, follow up research was presented answering some of the concerns of the directors from the first conference. The second conference also presented the recommendations of IGAD Parliamentarians and their video testimony on the importance of investing in conservation as a livelihood strategy. Although follow up on the initial recommendations by the directors themselves was not at the level expected, the conference agreed to further recommendations, which are being taken forward by a regional Trans-boundary Biodiversity Management initiative led by IGAD and the European Commission.

1996). At this level, well-informed experts can participate in influence debates that lead to changes that modify the overall thrust of debate on a given subject. By altering the opinion of decision-makers, experts can potentially change the range of possible policy agendas and outcomes.

Governments often have to respond to short-term political issues, for example, following up on election promises or responding to arising emergencies, for which they rely heavily on their Public Servant experts. For experts this may create an opportunity to push for policies that would not only help the government in achieving their agendas but also to reform existing policies. In most African governments, government experts are instrumental in ensuring that governments' short-term agendas are achieved (James, 1993).

External agents may struggle to respond to such short-term opportunities and therefore need to maintain communication with government experts to ensure that they are forearmed with facts and policy advice. By routinely equipping government experts with facts and advice it has been possible to highlight the importance and magnitude of particular issues giving them increased attention in terms of resource allocation and development of policy agendas (Barkenbus, 1998).

Nevertheless, it should be noted that researchers have often struggled to translate their policy advice into policy agenda-setting. Strong research has not contributed to significant restructuring of environmental priorities in most African countries and researchers need to examine their approach to engaging with government experts to ensure that their research is disseminated at the right levels and right stage of policy deliberations. Improved communication is particularly important

FIGURE 5.2 Educating communities to make their case to government, Garba Tula, Kenya (IUCN, 2010)

to ensure that advice is packaged in a way that is understandable and practicable (Davies and Mazurek, 1998). This issue is picked up and explored in depth in Chapter 8 of this book.

Influencing the policy formulation process

Once an issue reaches the policy-making agenda, government experts are usually called upon to identify options and deliberate on the most appropriate solution. Ideally, at this stage, a range of options should be considered and deliberated upon and the best solutions should be sought. Different policies may require different approaches to determining the best solution and therefore this stage of the policy-making process does not follow a set of blueprints or methodologies (Barkenbus, 1998). It is also likely that this stage of policy-making will come under political influences pushing for a particular option or policy.

For politicians to be able to fulfil election promises, they may call upon experts to propose credible options. The pre-determined agenda may demand particular methods and approaches that require scientific backing and legitimacy: a case of policy driving science rather than vice versa, as discussed in Chapter 1 and outlined further in Chapter 7 in this book. Politicians may, for example, require technical advisors to render policy choices credible by fortifying them with expert knowledge, methodological applications and reasoned expectations (Khoo, 2010). In the case of the Uganda Wetlands Policy, presented in Box 5.2, the government made a firm commitment to ensuring the sustainable management of the countries extensive wetlands and charged the Ministry of Water, Lands and Environment to work with international development partners to develop the necessary policy and legislation.

Through research, technical advisors are also able to identify shortcomings of current policies and give suggestions on appropriate policies and how best they can address societal problems. They are therefore able to contribute to the analysis of proposed policies by determining both their strengths and weaknesses and the possibility of the proposed policies being effective once implemented. A policy body like KIPPRA in Kenya has been instrumental in conducting research and policy analysis in all sectors of the Kenyan economy, providing capacity building for policy-making and implementation, and has also served as a point of contact for exchange of views on public policy issues affecting Kenya (www.kippra.org/content/7/2/about-kippra#).

In the policy formulation stage, government technical advisors analyze the implications of different courses of action in solving identified problems or issues. Analyzing and assessing different courses of action requires detailed knowledge of how a particular area of policy works and therefore experts with a more specialist focus are engaged. Multi-disciplinary experts have also used this approach of influencing policy formulation. Engaging technical advisors who are well informed of the specific policy area is particularly useful if there is resistance to

Box 5.2 *Uganda Wetlands Policy*

The process of developing and implementing Uganda's Wetlands Policy illustrates the many stages that a policy goes through. The political push for this policy from the 1980s led to the early ratification of the Ramsar Convention in 1988 and was followed by adoption of the National Wetlands Policy in 1994. The policy process was led by the National Wetlands Programme launched in 1989: a two year project that eventually led to formation of a fully-fledged Wetlands Inspection Division within the Ministry of Water, Lands and Environment. After adoption of the Wetlands Policy, wetlands were written into the constitution and subsequently adopted in a number of laws, including the Local Government Act and the Land Act (Mafabi *et al.*, 2005).

The policy process was supported through knowledge management on Uganda's wetlands, including a National Wetlands Inventory and setting up of the Uganda National Wetlands Information System. The process was strengthened by a strong consultative process that led towards policy formulation. Consultations and studies highlighted the absence of wetlands-related legislation in Uganda, and were very influential in drafting the policy. The policy itself was a critical step to guide subsequent legislation, recognizing that wetlands fall under numerous ministries and policies and therefore no single piece of legislation would be adequate (Ramsar Convention Secretariat, 2010).

an initiative from vested interests claiming that a particular policy cannot work (James, 1993).

Government technical advisors can influence the policy formulation process by enlightening policy-makers and therefore changing their general climate of thinking. This can be done through continued, but perceptible, altering of how decision-makers perceive issues and their solutions (Oh, 1996). Technical experts can help to correct informational biases in policy-making processes that would otherwise lead policy-makers to generate inadequate policies. Government technical advisors with no ties to special interests can on the other hand provide a broader perspective on policy issues, often bringing with them a wealth of experience and a set of policy options unconstrained by narrow interests. The enlightenment value of expertise, therefore, can in some instances increase policy space and enhance the potential of formulating unbiased win-win policies. On the other hand, policy-makers could also payoff technical experts to support their interests, which would compromise the legitimacy of policies formulated and hence science over complex issues could be twisted to support certain viewpoints.

Government technical advisors also use research in identification and formulation of policy objectives, which are instrumental in the policy formulation stage of the policy processes. Through adequate research, they are able to speak with authority on the choice of policy objectives since they are able to determine whether certain objectives, if attained, would contribute to the ultimate goals sort by politicians (Lindvall, 2006). For instance, if politicians are seeking office, they will often have a stronger motive to listen to experts who are able to estimate how much of a threat a certain policy outcome, such as high unemployment or increased poverty levels would be to their electoral prospects.

Government technical advisors do not only draw their advice from science but can also be crucial in bringing non-scientific perspectives into policy debate: for example, the views of communities. This is an important role since political leaders in most cases also need to combine the expertise of technical experts with the wisdom and intelligence of other individuals as well as those who are important to their popularity and success. By combining the knowledge of a particular subject area with the qualities possessed by these key individuals, politicians are able to make enlightened decisions that would not be entirely to their own benefit but also benefit their electorate (Barkenbus, 1998).

While the role of government technical advisors is instrumental for determining different policy options and giving scientific backing and legitimacy, it should be noted that a host of other factors collectively influence the policy-maker in making policy choices. These would include for instance the electorate and donors. Therefore, while considering the role of government technical advisors in the formulation stage, it would be important to also involve the electorate since they are the ultimate beneficiary of the policies. Donors also play an important role in policy formulation but they can work in collaboration with technical advisors in ensuring that the right policies are established (Barkenbus, 1998).

Influencing policy implementation

Policies on their own are not self-executing and to deliver results they need to be implemented. As the Uganda Wetlands Policy example (Box 5.2) illustrated, policies can set an agenda that can then be picked up in one or more legislations. To ensure achievement of positive results from formulated policies, adequate implementation measures are required (Ingram and Mann, 1980). Government technical advisors are frequently called on to develop policy instruments to enable implementation.

Government technical staff may play a significant role in the development of policy instruments. Their institutions are charged with policy implementation and they have responsibility for turning policy into practice. Public institutions may commission research to support the legislative process, and scientific backing is often sought to give legitimacy to legislation. For a policy to achieve its intended objective, it must be able to be implemented and adjusted to the conditions on

FIGURE 5.3 Farmers and extension workers experimenting with sustainable farming techniques, Mount Elgon, Uganda (IUCN Uganda, 2010)

Box 5.3 *Tanga Coastal Zone Conservation and Development Programme*

The Tanga Coastal Zone Conservation and Development Programme (TCZCDP), as presented in Chapter 2, ran from 1994 to 2007 and developed approaches for adaptive co-management of Marine Protected Areas. The role of Government Experts was crucial to the success of this programme and to the continued support for Collaborative Management Areas (CMA) in Tanzania. The TCZCDP engaged government experts from the outset and provided training and capacity building during the programme and close involvement throughout implementation. As a result, government technical staff were fully aware of the innovative approaches being used, were well placed to identify policy gaps or bottlenecks, and were more motivated to tackle policy constraints.

The success of this initiative, and the ongoing support for CMAs in Tanzania, was attributed to the team spirit that developed through the programme. In particular the programme effectively developed collaborative action planning and clear engagement with government and communities together, based on communication and respect. "What made it successful? District officers working together as a team will continue. A cadre has been built, and this is how they now do business" (Soloman Makoloweka, Tanga Regional Fisheries Advisor, cited in Wells *et al.*, 2007).

the ground where it is being implemented. It is therefore the role of government technical advisors to ensure that policies developed are able to be implemented so as to achieve the set objectives.

Government technical advisors from the national to local government levels are charged with implementation of government policies, and often are expected to identify the most appropriate means of implementation. Some policy decisions may be made in the form of Executive Presidential Directives or similar, and public servants are expected to accommodate them within the resource envelope and methodology of implementation. Such decrees may be based on vested interests or may be a reaction by the President or the Executive to crises and emergencies. Public servants are expected to rubber stamp such directives and provide the means for their realization (Oleh, 2004).

Government technical advisors are also instrumental in providing clarification for unclearly worded policies or legislative mandates (Barkenbus, 1998). For instance, in polices advocating for 'safe' levels of pollution or 'clean' production processes technical experts are expected to provide science-based determination of what are acceptable targets as well as adequate explanation of the meaning of these phrases to the public as well as to politicians.

Utilization of the expertise of government technical experts in this phase of the policy processes is both for standard setting of policies and for technical assistance. Laws and policies are frequently made at high levels, far above the level at which they will be implemented, and therefore technical assistance is often

Box 5.4 *Control of* Parthenium hysterophorus *in Kenya*

In 2010 IUCN and CABI Africa, as partners in the Global Invasive Species Programme, prepared a Briefing Note for the Kenyan Minister of Agriculture on the importance of the invasive species *Parthenium hysterophorus*. The note recommended declaring *P. hysterophorus* a noxious weed in Kenya because of its impact on agriculture, human health and livelihoods and on the environment and biodiversity. It was further recommended to publicize this decision, to develop management strategies to contain the spread of this weed and to eradicate isolated infestations, and in the long-term to explore host-specific biological control agents.

The Briefing Note was shared with the Kenya Plant Health Inspectorate Services (KEPHIS), which lies in the Ministry of Agriculture. KEPHIS successfully raised the issue with the Minster and on 13 April 2010 the Minister formally declared *P. hysterophorus* a noxious weed under the Noxious Weeds Act (Gazette Notice No. 4423: The Suppression of Noxious Weeds Act (Cap. 325)).

needed to assist local communities in determining how best to enact policy measures. The consistency and insight of this assistance is therefore critical to the accomplishment of policy goals (Barkenbus, 1998).

Influencing policy evaluation

Policy evaluation is in most cases a forgotten part of the policy cycle (Davies and Mazurek, 1998). Priority has instead been put on the creation of new policies than evaluation of existing policies (Portney, 1998). It should however be noted that policy evaluation provides feedback to policy-makers, which can be incorporated in future decision-making and policy formulation processes.

The infrequency of policy evaluation could be attributed to the fact that in the political context it might prove embarrassing to those responsible for policy

Box 5.5 *Evaluating bottlenecks in environmental policy in Tanzania*

As part of the process of developing the General Management Plan for Tanga Marine Protected Area, research was conducted into the policy environment, particularly to identify bottlenecks in environmental policy. The research was designed to identify factors hindering effective formulation and implementation of environmental and development work in coastal areas in Tanzania and to inform livelihood development activities for the Tanga Coelacanth Marine Park (Mwaipopo, 2010).

Factors that were identified include:

- Inadequate conceptualization of livelihoods to allow policies to capture the different dispositions of people in the context of conservation.
- Conservation adopting resource-oriented approaches rather than people-oriented approaches.
- Unclear property rights over resources.
- Limitations in the devolution of powers.
- Inadequately mandated or evolved community-based organizational capacities to handle and stimulate positive livelihood and resource management activities.
- Limited strategies to assemble more meaningful stakeholder collaboration in order to support livelihoods.

These recommendations were adopted during the formulation of the General Management Plan to ensure that constraints were mitigated, or were targeted for policy reform.

formulation and implementation. For the same reason, policy evaluation may be favoured by opposition parties and can be carried out to score political mileage. It should also be considered that decision-makers function in a political environment where the success of a policy is critical for their re-election. Therefore if it is perceived that the evaluation of a policy could produce negative results, such evaluations may be ignored, whereas if the opposite is true, policy-makers are more likely to encourage policy evaluation (Barkenbus, 1998).

Although policies may not be routinely evaluated by government, informal evaluations take place all the time, by technical experts, the media and other organizations (interest groups). Institutions charged with the implementation of policies have a vested interest in policy evaluation to ensure that their objectives for the policy are assessed. With the increasing interest in performance-based monitoring, there is increasing attention to policy evaluation (Barkenbus, 1998). For instance, levels of children enrolling for the free primary education in Kenya are being used as an indicator for either the success or failure of the free primary education initiative.

Assumptions made when working with government technical advisors to influence policy processes

Government technical advisors are highly knowledgeable and experienced in their different fields

The main assumption made when engaging experts is that they are highly knowledgeable in their field of expertise and that they can offer solutions to societal problems. In countries with a high level of democracy the level of trust in technical experts is often quite high, especially in finding solutions to societal problems (Iwona *et al.*, 2002). In these countries these authors note that up to 92 per cent of the population would prefer the policy-making processes and decision-making to be left to the judgment of experts. The trust in experts according to these authors is greatest in countries where the social and economic problems are enormous and therefore experts are viewed as the ultimate resort to improve the situation.

Government technical advisors are expected to be highly knowledgeable and therefore a great resource in the policy formulation process since they are able to carry out adequate research and give advice based on credible and unbiased information. This assumption about government technical advisors can however be compromised especially if government technical advisors are appointed to their positions by politicians with their own interest at heart. For instance, in appointment of permanent secretaries and directors of government parastatals including research organizations, the President or Minister may appoint their cronies rather than make appointments based on merit. This greatly affects the ability of such government technical advisors to perform their duties, both

Box 5.6 *Promoting Beach Management Units in Eastern Africa*

Between 2002 and 2005 IUCN worked with the Lake Victoria Fisheries Organization (LVFO) to address environmental concerns and sustainable management around Lake Victoria. The Lake is shared by three countries: Kenya, Uganda and Tanzania, all of which make provisions in their domestic policies for devolving management responsibilities to local groups. These groups are collectively referred to as Beach Management Units or BMUs.

IUCN and LVFO partnered with the respective fisheries departments and fisheries research institutes in three partner countries. This included the Kenya Marine and Fisheries Research Institute (KEMFRI), the Tanzania Fisheries Research Institute (TAFRI) and the Fisheries Resources Research Institute of Uganda (FIRRI).

Between 2003 and 2004, IUCN and LVFO facilitated a series of initiatives to raise awareness of the relevant policies and the LVFO Convention and to build motivation and capacity to develop BMUs. Working with local and regional government officials and local communities the project developed communications and technical papers and organized cross-border exchange visits to promote sharing and learning. The partners also organized an international workshop in October 2003 and a publication entitled *The International Workshop on Community Participation in Fisheries Management on Lake Victoria*. The workshop brought together BMU members, district and regional fisheries officers, and senior members of the Fisheries Research Institutes and Fisheries Departments from the three Partner States as well as representatives from international institutions and programmes working on issues of co-management in fisheries. The workshop provided a review and analysis of BMU development on Lake Victoria, and it contributed towards the development of a common vision for a co-management regime in Lake Victoria.

Based on these experiences, the three countries have scaled up their support for BMUs and have replicated the approach in other parts of the region, including coastal areas as well as fresh water lakes. BMUs now receive support from their respective governments and are registered with their line Ministry responsible for fisheries. Additionally, recommendations from the project, published in a 2004 report on cross-border fishing and fish trade were adopted by the Council of Ministers of LVFO and became a Ministerial Council decision to guide fishing and fish trade on Lake Victoria.

because they may lack the required expertise and because they are likely to be more prone to political manipulation by their patron. The technical capacity of Public Servants can also be called into question where there are high levels of corruption in the public service and where civil servants are frequently bribed by external agencies to influence the technical advice they are dispensing.

Government technical advisors are able to influence the incorporation of issues raised in policy debates at different government levels

When working with government technical advisors in trying to influence policy, it is also assumed that they will be able to raise issues raised to their national governments or at regional forums. Policy discussions with technical advisors are expected to be fed back to their national governments and contribute to policy reforms or even creation of new policies. This assumption about government technical advisors is however not likely to hold if their governments have other urgent issues to address, like ensuring peace and security and rebuilding a country after civil strife. The assumption may also not hold if the technical advisors involved are not influential at national and regional levels. It is therefore important to incorporate country goals and objectives when engaging government technical advisors in trying to influence policy. This will prevent parallel objectives for both the government and the organization engaging technical experts.

Government technical advisors are non-partisan

When engaging experts in policy processes, it is also expected that experts are non-partisan and that they would give unbiased judgment and opinion regarding policy issues. Their judgment is therefore always expected to be objective and based on factual data other than being based on political backing and influence. Separating the role of government technical advisors from the influence of politicians is however a challenge especially given that some of the government technical advisors are appointed to their positions by politicians and not by merit, as discussed above. Government technical advisors appointed by political figures include permanent secretaries and parastatal heads. This may therefore have an impact on the ability of the government technical advisors to carry out their duties without political bias.

Challenges of involving technical experts in policy processes

Lack of political resources to adequately influence policies

Policy formulation processes in most African countries are considered to be political processes that require political resources and backing for the policies to be agreed on and implemented. Scientific or technological knowledge on its own is not considered sufficient to influence the policy process (Brint, 1994). Therefore, for government technical advisors to be able to adequately influence the policy processes, they require political resources that are, in most cases, vested in the politicians. Major events other than the technical contribution of technical experts to policy process are also likely to influence the policy formulation process by affecting the opinions and views of policy-makers (Nelkin, 1992).

Technical advisors may also be limited in their ability to make independent judgments based on their technical expertise, except in specific technical matters, since their appointment could be politically motivated (Brint, 1994). Politicians on the other hand have the political resources and mandates to influence the larger public and the ability to ensure that policies are formulated and implemented. This has tended to limit the degree to which the ideas of government technical advisors are able to influence policy since they will require the backing of politicians, which they are only likely to get if the policy favours the politicians interests and objectives. If the policy is contrary to the interests of the politicians, then it is most likely that it will not see the light of day.

Dilemmas for democratic regimes

Use of government technical advisors in policy processes can cause problems for democratic governments (Corrales, 2002). This is because by using the technical advisors entirely in the policy process the plurality of opinions, which is essential in most democratic states, is ignored (Centeno, 1998). Technical advisors, being highly knowledgeable in their respective fields, may interact only with colleagues in their profession and exclude or lock out other potential stakeholders in the policy process. This could cause the "common man" to feel detached from the government's policy processes and lead to public distrust of certain policies.

By ignoring the role of the public in policy processes, policy dialogue is restricted and structures of accountability become distorted. This can happen, for example, if governments become more accountable to multilateral agencies and donors than to their own institutions and the populations. Furthermore, over-emphasis on technical advisors can lead to breakdown in communication, since advisors may use highly technical language to justify policies, leaving Civil Society and the public unaware of what is going on (Conaghan, 1996).

Undermining the legislature

Continuous use of government technical advisors in the policy processes may undermine the legislature. The legislature is always charged with passing of laws in a country but with the rise in the use of technical experts in policy processes, then the role of the legislature could be ignored or side-lined. On the other hand, if the legislature feels that they are being side-lined by the technical advisors, they are likely not to pass policies formulated by them when they get to parliament. This would result in a push and pull between the technical advisors who are meant to advise the legislature and the policy-makers who constitute the legislature who pass the policies (O'Donnell, 1994).

Alternatively, the rise in use of government technical advisors could increase the decision-making capacity of policy-makers. By continuously seeking the advice of technical experts, policy-makers can become better informed and more capable of contributing to policy debates (Corrales, 1997).

Nature of a country's political reforms

Political reforms in a country are likely to affect the ability of government technical advisors to influence the policy processes. If the political reforms are in support of devolution of power to professionals with adequate skills, then the role of technical advisors will be valued and strengthened, but if the opposite is true and decision-making power is centralized, the role of experts is more likely to be ignored. This is likely to lead to unrest and opposition of government policies by the Civil Society and the public (Bangura, 1994).

Conclusion

The increasing complexity of societal problems and dilemmas calls for adequate scientific knowledge and expertise in order to come up with solutions that deal with these problems. Whatever solution is decided upon to address the problems, it should have adequate scientific backing. Engaging experts with scientific knowledge and expertise is important in ensuring that viable solutions are sought to address societal problems.

The role of government technical advisors in the policy formulation process can therefore not be overlooked since they can be highly experienced in their different disciplines and their involvement in the process is therefore significant. However, for them to be effective in the policy formulation process there is need to separate their work from political influence and to ensure that they incorporate the views of all stakeholders more especially the citizens that are to benefit from the policies.

NGOs and other development organizations seeking to influence policy by working with government technical advisors need to build relationships with government technical advisors and involve them in their work from the onset. When planning to work with government technical advisors to influence policy, it is important to have a clear idea of what exactly the government technical advisors can influence at the different levels of government.

To have an impact on policy processes, non-governmental organizations should consider working with government technical advisors to influence the government's medium- to short-term agendas, since this seems not to have been well exploited. This would require organizations to be up-to-date with political events and priorities at national and regional levels.

The role of government technical advisors in the policy process is likely to be felt more in development of policy instruments than policy goals, since the latter are more likely to be determined by politicians. Therefore, when engaging with government technical advisors, organizations should develop a clear idea of what they are able to influence in the policy process and what they are not able to influence. Above all, external agents should develop a good understanding of the actors, processes and policies that they are trying to influence, and an understanding of where political patronage is at play and which audiences will listen to which messages.

6

INFLUENCE WITHOUT BORDERS

Regional approaches to environmental policy dialogue

Barbara Nakangu Bugembe

Introduction

In June 1992, the Rio Earth Summit declared that "the right to development must be fulfilled so as to equitably meet developmental and environmental needs of present and future generations". This was based on the Brundtland Report that raised increasing concerns about the effect of economic development on health, natural resources and the environment (World Commission on Environment and Development, 1987). It clarified that sustainable development was not just about the environment, but also the economy and the society. The summit produced a comprehensive blueprint of action, known as Agenda 21, to be taken globally, nationally and locally by organizations of the UN, governments, and major groups in every area in which humans directly affect the environment. The 178 countries that signed Agenda 21 were encouraged to implement sustainable development from the local to national to regional levels.

Many nations have since made efforts towards fulfilling their commitments to this agreement albeit with challenges especially in developing countries. A 1998 review found limited or no evidence of successful implementation of the environmental mainstreaming concept in developing countries, and environmental outcomes remained as mission-statements, or long-term goals with no specific action plans or budget allocations for their implementation. At the time this was attributed to poor understanding of the concept, limited capacity to undertake mainstreaming and simplistic expectations that "off-the-shelf" mainstreaming tools, such as the Strategic Environmental Assessments tool (SEA), would be readily adopted (Brown and Tomerini, 1998). More recently efforts have been undertaken to rectify this, such as the joint UNEP-UNDP Poverty Environment Initiative.[1] Nevertheless, such mainstreaming work is carried out at national level

FIGURE 6.1 Mount Elgon in Uganda is part of the Nile Basin and environmental management has profound downstream implications for neighbouring countries (IUCN Uganda, 2010)

and the role of Intergovernmental Bodies, and the best way to engage with them for the sake of influencing policies within their member states, remains uncertain.

This chapter explores approaches to working with Intergovernmental Bodies (IGBs) to mainstream environment in development policy. The chapter illustrates that working through IGBs can be an important avenue for influencing environmental policy. Further, it also asserts that this approach needs to be considered carefully based on clear objectives as it can be time consuming, expensive and may not lead to actual implementation on the ground. IGBs have different structures and mandates and these need to be understood in order to identify appropriate ways of engagement.

The chapter presents a framework that describes conditions to consider in engaging IGBs in policy influencing. It builds on the description of policy presented in Chapter 1 that the policy process should be viewed as a course of action rather than an end in itself. It emphasizes that the focus should be on knowing what to influence and determining the best course of action to take.

The time to share these lessons is opportune given that the global and multidimensional character of environmental problems and most countries accept that holistic international efforts are required to address many aspects of environmental deterioration (Hart, 2008). Regional organizations provide the opportunity to engage collectively to influence environmental policy and implementation. They allow for broader representation and peaceful persuasion through norm sharing and complementary institutions that can converge to affect the collective objectives of the countries, Moore (2004).

The chapter draws lessons and experiences from initiatives in Eastern and Southern Africa that have worked to influence policy through regional bodies. Specifically the IDRC supported programme known as Conservation as Assets for Livelihoods,[2] and the Norwegian funded programme known as the Mount Elgon Regional Conservation Programme (MERECP).[3] These were used to engage regional and intergovernmental bodies in influencing conservation policies targeting national, sectoral planning and budget processes.

While the entry point for policy influence could be through IGBs, the main objective is to achieve environmental mainstreaming in sectoral plans and budget processes at the national level. The Poverty and Environment Initiative (PEI) mainstreaming guidance note (UNDP and UNEP, 2007) emphasizes targeting existing country planning activities for mainstreaming the environment. It further notes that national and sectoral planning and budget processes are best for establishing enduring institutional processes that can integrate environment into planning and decision-making within government, particularly line ministries responsible for sectoral infrastructure and growth.

The nature of intergovernmental or regional bodies (IGBs)

Regional or intergovernmental bodies are institutions created when states enter into agreement to cooperate on a number of issues. The objectives are often economic, political or security related, however these may be expanded to include natural resources management, climate change and migration among many others (Moore, 2004). The agreement to cooperate is sometimes referred to as integration. It may also refer to agreeing on a criterion to achieve a specific goal, such as collaborating on management of a transboundary natural resource, or establishing an institution to coordinate the implementation of some of the provisions of the agreement – a regional body. Most industrial and developing countries in the world are members of regional integration agreements, and many belong to more than one.

The "depth" of integration, is the level of collaboration countries have agreed on. Some IGBs work on loose collaborations such as simple removal of tariffs or a sectoral agreement to manage a trans-boundary resource while others aim at more areas of collaboration. The level of collaboration is dependent on the objectives and constraints of the countries that are involved in the cooperation. For some countries, the objectives are to secure economic benefits as well as build deep political links, and there is a willingness to exert political effort and establish institutions to meet these objectives. In this case they aim for deep integration, a kind that may bring larger benefits in many areas. However, deep integration also involves greater political commitment, and far more complex policy-making than would loose sectoral based agreements. The level of integration determines the institutions that are formed and the types of activities that are carried out, which determines how the institutions can be engaged to influence policy. Most countries that aim to cooperate usually undertake a phased approach to achieve full integration with indicative milestones and timelines.

The East African Community (EAC)[4] is an example of a deep regional integration. From the start the EAC set out to achieve a political federation but has phased this to achieve various milestones along the way such as the customs union, common market and monetary union. Another approach is that countries may start with a single cooperation objective but then through negotiations build on it and progress along the continuum to include many other aspects of mutual interest. This was the case of Southern African Development Community (SADC) that initially cooperated to achieve political liberation of Southern African countries and when this objective was achieved it proceeded to include promotion of economic and social growth, and achievement of peace and security.[5] Consequently, it is important to evaluate and understand each IGB being engaged. Issues such as mandates, goals, plans and structures need to be compared to determine if the institution can be used in the policy influencing effort being pursued.

Examples of intergovernmental bodies in East Africa

The East African Community (EAC)

The EAC is a regional intergovernmental organization that was established by treaty in 1999 and entered into force in 2000 by the Republics of Kenya, Uganda and Tanzania. Rwanda and Burundi later joined in 2007. The organization was formed to widen and deepen cooperation in political, economic, and social aspects of the member countries with the ultimate goal of achieving a political federation. However, the EAC is pursuing a phased approach towards this goal and has set milestones that include establishing a Customs Union in 2005, a Common Market in 2010, a Monetary Union by 2012 and ultimately a Political Federation of the East African States. Operationally, the EAC structure includes a policy organ and a number of institutions that coordinate day-to-day operations.

The policy organ is composed of the Summit, Council of Ministers, Coordination Committees, Sectoral Committees, Court of Justice and the East African Legislative Assembly. The EAC institutions include a Regional Secretariat located in Arusha, Tanzania; the Lake Victoria Basin Commission located in Kisumu, Kenya; The Lake Victoria Fisheries organization located in Jinja Uganda; the East African Development Bank with it regional offices in Kampala, Uganda but with country offices in the member countries; the Inter-University Council of East Africa, whose regional secretariat is in Kampala, Uganda and the Civil Aviation Security and Oversight Agency located at Entebbe, Uganda.

The EAC is guided by a development strategy to facilitate the implementation of the treaty in a systematic manner. This has been reviewed three times since 2000. It is also important to note that each of the specialized institutions of the EAC have specific strategies and plans that guide their implementation (EAC, 2011). The structures and processes of the EAC indicate that it aims for "deep" integration and thus has established commensurate institutions. The different levels of the policy structure as well as the different institutions provide various entry points for policy influence including the highest political levels.

The Nile Basin Initiative

The Nile Basin Initiative (NBI) was initiated through dialogue among the ten Nile riparian countries that resulted in the agreement to collaborate to achieve a shared vision – to "achieve sustainable socioeconomic development through the equitable utilization of, and benefit from, the common Nile Basin water resources". The NBI was set up without a regional or international treaty or agreement among the riparian states. However, there was agreement to cooperate under a "Transitional Institutional Mechanism of the Nile Basin Initiative (NBI)", that would inform the targeted "Cooperative framework agreement".[6] Contrary to the EAC, the operational structure of the NBI includes the Council of Ministers of Water Affairs of the Nile Basin Countries (Nile-COM), as the highest policy organ, which provides policy guidance and makes decisions on matters relating to the agreement. The Chair of the Nile-COM is shared among the member states on an annual rotational basis. The Nile-COM is informed by the Technical Advisory Committee (Nile-TAC), which provides it with technical advice and assistance. The Nile Basin Secretariat (Nile-SEC), located in Entebbe, Uganda renders administrative services to the Nile-COM and Nile-TAC.

The Intergovernmental Authority on Development

The Intergovernmental Authority on Development (IGAD) encompasses seven countries of the Greater Horn of Africa: Somalia, Sudan, Ethiopia, Djibouti, Kenya, Uganda and Eritrea. IGAD was created in 1996 to supersede the Intergovernmental Authority on Drought and Development (IGADD) which was founded in 1986. The institution was initially created in response to recurrent drought in this largely dryland region but has since been expanded to include aspects of economic development and security. The IGAD is guided by a Sub-Regional Action Programme (SRAP) for the implementation of the United Nations Convention to Combat Desertification that was published in June 1998 after four years of consultation.

The IGAD secretariat is located in Djibouti city, Republic of Djibouti. Its operational structure includes the Assembly of Heads of State, which is the supreme policy-making organ and determines its objectives, guidelines and programmes.

It meets once a year and is chaired by an elected member state on rotation basis. The Council of Ministers composed of the Ministers of Foreign Affairs and one other Focal Minister designated by each member state formulates policy, approves the work programme and annual budget of the Secretariat during its biannual sessions. The Council of ministers is informed by the committee of Ambassadors, which is comprised of IGAD member states' Ambassadors or Plenipotentiaries accredited to the country of IGAD Headquarters. It convenes as often as the need arises to advise and guide the Executive Secretary who heads the secretariat.

The Secretariat assists member states in formulating regional projects, facilitates the coordination and harmonization of development policies, mobilizes resources to implement regional projects and programmes approved by the Council and reinforces national infrastructures necessary for implementing regional projects and policies. The Executive Secretary is assisted by four Directors heading Divisions of Economic Cooperation & Social Development, Agriculture and Environment, Peace and Security, and Administration and Finance.[7] Just as the EAC, the IGAD provides opportunity to engage the highest political organs. However, it has fewer institutions and thus a narrower range of entry points for policy influence.

Role of intergovernmental bodies

The general understanding of most IGBs is that they are formed for economic, political and security reasons. The advantages often include reducing costs of production and transaction, enhancing a countries competitive advantage, ensuring social cultural inclusion and security. However just as the setup of various IGBs differs, so does their ability to achieve certain objectives. There is a need to undertake a good analysis of the target IGB's likelihood to contribute towards achieving a required policy objective. If some of the roles indicated below do not contribute to the attainment of the intended objective, then the strategy to engage IGBs may need to be reconsidered.

IGBs enhance trust and security for regional members

The OECD paper on regional blocks argues that the primary role and driving factor of regional integration is enhancing political benefits of the member countries. Other benefits such as economic, social and environmental are the effects of the political payoffs. It is further argued that the key political benefit of IGBs is enhancing regional security. This may result from the ties of collaboration and frequent policy level contacts during regional negotiations that raise the degree of trust among participating nations. Additionally, the enhanced benefits from cooperation may make conflict expensive.

Enhanced trust and security may lead to greater consideration of issues of mutual interest such as natural resource benefit-sharing. The Mount Elgon Transboundary

Box 6.1 ***The MERECP Project*[8]**

The Norwegian funded Mount Elgon Transboundary Project (MERECP) implemented between Uganda and Kenya was considered and approved at a scheduled summit meeting of high level policy-makers due to the trust that had been cultivated over the years of negotiating the various regional issues. It was highlighted that the project would support the achievement of a number of EAC regional strategic plans. For example, the project would improve ecosystem security by curbing illegal activities in the area that were exploiting the freedom of cross border movement and the lack of regulation over resource management.

Project, implemented in partnership with the East African Community, enabled negotiations that led to trans-boundary environmental concerns being embedded in local development plans (Box 6.1). It is important however, to evaluate the political achievements that would attract attention and at the same time provide environmental benefits when considering working through IGBs to influence policy.

Enhance the bargaining power and synergies of member countries

IGBs may enhance the collective bargaining power of the member countries. However this depends on the ability of the member countries' agreement on a common position during negotiations – an aspect that has largely eluded IGBs from developing countries (Moore, 2004). Nevertheless, IGBs in developing countries have supported a number of their member countries to raise attention of particular key issues though research, knowledge sharing, collating and consolidating statements for presentations at summits and ministerial levels (AU, 2005).

In 2007 a meeting involving high level technical advisors from the finance and environment sectors of the six IGAD Partner States was organized to raise awareness about the value of natural resources (IUCN, 2007b). Subsequently, the IGAD Director of Agriculture and Environment took advantage of the knowledge and interest created at the meeting to follow up and engage the technical officers to develop a transboundary natural resources management programme. The programme was based on the enhanced awareness and appreciation of the unique common issues affecting the IGAD countries as had been identified during the meeting, as well as the participants appreciation that there was a need to collaborate on a common programme that captured their synergies (see Box 6.2).

Box 6.2 *Outcomes of the 2007 Conference of IGAD Directors of Conservation, Finance and Planning*

The October 2007 Conference of IGAD Directors of Conservation and Directors of Finance and Planning on "The Environment and Natural Resources as Core Assets for Sustainable Development" made the following recommendations for consideration by the IGAD Council of Ministers:

1. The IGAD Sub-Regional Environment Action Plan (SREAP), the IGAD Sub-Regional Action Plan (SRAP), the IGAD Environment Outlook (IEO) and the IGAD Environment and Natural Resources Strategy should all be implemented.
2. IGAD should create an ad hoc task force from the region to address important environmental issues including briefing sessions for Members of Parliament from the region.
3. IGAD should initiate a process towards a common regional approach on the governance of natural resources as a core asset and of strategic value as well as being a part of the African Peace and Security Architecture.
4. IGAD should conduct a regional inventory of the biophysical, socio-economic and cultural assets of, especially, the arid and semi-arid lands, which is the dominant ecosystem in the IGAD region. Knowing what the region has will allow appropriate governance systems, marketing channels, resource valuations, etc. to be put in place.
5. Concurrently with the inventory of the ASALs, products from underexploited natural resource should be identified, analyzed and pre-feasibility studies undertaken for their commercialization in national and international markets.
6. The IGAD Secretariat and its partners should develop the tools and methodologies to capture the true value of the environment and natural resources.
7. IGAD should foster the compilation and sharing of research and technology information available in the region.

Support member countries to commit to regional and international obligations

Designs of many regional agreements provide commitment mechanisms to ensure that member countries implement international decisions they have agreed upon. This is useful as international agreements are generally voluntary and many lack serious sanctions in the case of non-compliance or poor performance. Thus IGBs usually make provisions to support in establishing common ground among their member states to enforce or implement these agreed decisions. Such initiatives include innovations to encourage the participation of sometimes reluctant States. An example is the Lake Victoria Fisheries Organization (LVFO) which was established to support the EAC and member countries in coordinating the management of the lake-wide fisheries resources (www.lvfo.org/index). The LVFO developed a uniform protocol and uses this to support member countries when undertaking regular research to monitor the management of fisheries resources and regularly presents these to the member countries to guide the formulation of measures for the sustainable exploitation of the resources.

Another example is the requirement that each member institution of the EAC presents progress reports about the implementation of agreed regional actions to the EAC policy organs (the sectoral council, council of ministers and the summit). This mechanism creates a sense of competition among the institutions and consequently encourages them to effectively fulfil their roles and responsibilities to present good outcomes and hence influence their funding and policy support. This was observed under the MERECP programme, where the Lake Victoria Basin Commissions (LVBC) was required to present progress about its implementation to the summit. LVBC ensured the enhanced delivery on interventions that needed to be implemented uniformly across the region since they were most important to report on at the policy meetings. As a result LVBC received political support in its efforts to put in place policies that enhance cross border activities such as patrols across the mountain. Specifically, LVBC received support to test out sensitive issues such as sharing radio frequencies that were necessary to implement cross border patrols (see LVBC sectoral reports on www.lvbcom.org/index).

IGBs mobilize and provide policy, financial support and technical expertise

This enhances cooperation and synergies among member countries. For example, the EAC and the Nile Basin Initiative have established basket funds that a number of donors contribute to and member countries draw upon to implement regional programmes. These have enabled member countries to undertake a number of important programmes that would have been otherwise difficult to undertake. The NBI, for example, has been able to implement regional programmes including Kagera Transboundary Integrated Water Resources Management and Development project shared between Burundi, Tanzania, Rwanda and Uganda; the Mara Transboundary Integrated Water Resources Management and Development project in the Mara and the Serengeti national parks shared between Tanzania and Kenya; the Sio-Malaba-Malakisi Transboundary Integrated Water Resources Management and Development project shared by Kenya and Uganda; the Lake Edward and Lake Albert Fisheries Pilot Project in Uganda and the Democratic Republic of Congo. These initiatives are all intended to address transboundary environmental management and policy.

Regional bodies have also coordinated various studies about their areas of jurisdiction such as the Lake Victoria basin and Nile Basin resources, which highlight the various socioeconomic, biodiversity, investment opportunities and gaps of the respective regions. This process ensures reduced costs for individual members to acquire this information as well as highlight areas of synergy (LVBC, 2011; NBI, 2011). For example, the NBI is coordinating a programme known as Multi-Purpose Track where a series of studies are undertaken to inform the development of common analytical tools, and a plan for coordinated investments. Assessments undertaken include: Watershed Management Cooperative Regional Assessments, Power Trade Investment Project such as the Eastern Nile and the Rusumo power project, Irrigation and Drainage Cooperative Regional Assessment.[9]

Engaging IGBs in influencing policy and practice

Diagnosing the situation

The decision to engage regional or intergovernmental bodies in policy influencing must be based on clear and well understood objectives. As has been already indicated, the value is not in the description of a particular policy but in understanding what is the aim. Whether it is supporting policy formulation or supporting implementation, this very first step is crucial as it determines whether engaging IGBs is the best approach to achieve the intended objective.

Experience has shown that policy influence through engaging IGBs is time consuming and expensive and the approach should only be considered when the value is obvious. As already indicated, most IGBs' successes spin off from political objectives and thus a good evaluation should be made to identify other indirect opportunities that enhance the benefit of achieving the target policy goal. However, careful evaluation may need to be undertaken to ensure that the benefits justify the cost incurred. On the other hand, there are clear cross-border issues that may have to be dealt with internationally and may be best achieved through IGBs, regardless of the costs, such as insecurity, human and livestock diseases, pests and invasive species.

Further, it has been observed that engaging IGBs usually achieves more at a strategic level, which involves formulation, harmonization or accreditation of policies that provide a favourable environment to implement programmes at national or sectoral levels. Hence engaging IGBs usually results in very little actual implementation on the ground. Achievements at the national levels are generally more influenced by sovereignty priorities than regional issues. Therefore, an analysis needs to determine whether regional achievements contribute and are necessary to achieve target policy goals at the national level, otherwise resources may be spent and time wasted to implement activities that result in insignificant change on the ground.

Most policy influencing programmes undertaken through IGBs aim at two main objectives: promoting the harmonization of action in the member countries or providing a framework that mandates the member countries to implement a particular policy. However, as illustrated below, this process is not necessarily straightforward and has varying degrees of success.

From 2006 to 2009, IGAD partnered with IUCN to push for increased budgets and implementation of sustainable natural resources management programmes that achieve win-win situations for socio-economic development and environmental health. The partnership was based on the realization that most countries in the region undervalue the importance of drylands as well as their traditional management and thus undermine their contribution to people's livelihoods and a country's economic development. Most countries were underinvesting in pastoralism and allocating resources to alternative land use options that are less sustainable, (Niemi and Manyindo, 2010). The partnership worked through IGAD's institutional framework to highlight the key issues and to mandate member countries to implement agreed outcomes. IGAD's role was also to provide a basis to monitor and support adoption of agreed programmes or policies. Thus it was expected that the IGAD secretariat would also enhance continuity of the achievements of the programme by incorporating project outcomes in their policies or programmes.

When diagnosing a policy challenge and planning a response, it is also important to take a closer look at the targeted regional institutions and understand their systems and procedures, and how they carry out business. This will help determine the opportunities and challenges and hence the best entry points and alliances to consider in the policy influencing quest. Each IGB's structure and mandate is unique: for example, the highest organ of the NBI is a council of ministers, whereas for IGAD it is the summit, which involves heads of government participating in regular meetings. Hence, once decisions are made at IGAD meetings, countries would be obliged to implement them as policy directives, while NBI decisions are not final and representatives usually require time to return to their countries and consult before decisions that commit their countries are confirmed.

The MERECP programme that aimed at enhancing the joint management of the Mount Elgon cross border ecosystem can be used to elaborate the EAC processes further. The programme was developed after the realization that the implementation of separate conservation projects on either side of the mountain shared between Kenya and Uganda differed and could benefit from being implemented uniformly. Actions such as Collaborative Resources Management of the park resources between the communities and park authorities were promoted on the Uganda side, while the Kenya Park management policy implemented a strict reserve control approach that allowed no access for the communities. Furthermore, it was determined that there were some interventions that were not being exploited that would enhance the better management of the ecosystem. These included cross border tourism and patrols between Uganda and Kenya. Thus, the value of promoting the transboundary management of the ecosystem to enhance its value was obvious. However, it was a new concept that needed to be promoted in the two countries.

It was evaluated that the EAC was the best suited institution to coordinate the MERECP project and ensure that the national governments were supported to entrench the management of the uniform management of the ecosystem in their plans and budgets. The project aimed at supporting the EAC and national institutions acquire skills, systems and procedures that would encourage the transboundary management of the ecosystem. It was also evaluated that working through the EAC would facilitate a faster process to enhance collaboration

Box 6.3 *Diagnosis of the EAC – Indicating the opportunities and entry points for engaging EAC in the Transboundary Management of Mt Elgon*

The diagnosis revealed that the vision and strategy of the EAC for the sustainable development of Lake Victoria basin indicated the Mt Elgon catchment as a key area of intervention. Thus, the MERECP focus on Mt Elgon would directly contribute to the EAC priority programmes and thus provided a good opportunity to interest the EAC in the programme. Second, the EAC was in the process of finalizing the protocol on cooperation on management of East Africa's natural resources and so the MERECP transboundary programme, provided an opportunity to demonstrate the application of this protocol. Third, the EAC was also in the process of establishing the Lake Victoria Basin Commission (LVBC) and MERECP would be one of the first and major programmes of the LVBC at the time. Finally, the EAC procedures require that any large programmes such as the MERECP implemented through the EAC be subjected through the approval process that involves all the stages of the policy organs including the summit. This provided an opportunity to influence the institution up to the summit.

between the two countries than starting at the national levels or sectoral levels. Additionally a diagnosis had been undertaken and found that a number of opportunities and entry points (indicated in Box 6.3) would facilitate working through the EAC. The process of approval by the summit was also considered very important since it ensured that the programme and the intended objectives would be presented at the highest political levels and if approved would be considered policy directive that would have to be implemented and regularly reported on.

Identifying policy targets and alliances

It is important to determine the best entry points, based on the intended objective of the programme and the diagnosis undertaken. The analysis above should have identified processes, opportunities and challenges the institutions were facing, the individuals or exogenous factors influencing the processes, such as the political situation and timing to determine the best entry points. For example, during the design of MERECP, it was identified that the EAC at the time was understaffed and required support in designing programmes as well as presenting advice to the policy organs. It was also determined that the EAC was in the process of establishing a specialized institution to manage Lake Victoria – the Lake Victoria Basin Commission (LVBC) – which would benefit from having a tangible programme of work to coordinate. Technical staff in the EAC were supported to prepare the transboundary programme based on lessons from previous national projects and these were presented to the various technical and advisory committees. The programme was finally approved by the heads of three partner states at the time through their regular "Summit" meeting. Approving it at the highest level also entrenched a strong commitment by member states since this required regular updates about the programme to the summit.

Determine interventions and implementation

Once the objective is clear, targets and entry points are determined, interventions to be undertaken can then be developed. However, this is often one of the tasks that are not well planned. Many attempts to influence policy are implemented in a disconnected way, with a scattering of activities that are poorly coordinated within and between institutions. Unrealistic assumptions are made about how an activity will lead to policy change and much greater preparation is required to make activities more effective, but also to make expectations more realistic.

Common policy-influencing actions include preparing and distribution of policy briefs, brochures, awareness messages and meetings. However, it is important to note that these actions and others are tools that help in policy influence, and not the end goal in themselves. The challenge is in determining when and how to use these tools. It is even more challenging when the target for policy influence is at the regional level where policy targets are further removed, hence making the possibility of tapping into their processes difficult and expensive. Nonetheless, opportunities do present themselves and to be able to exploit these opportunities it is important to remain abreast of ongoing policy processes and to maintain relationships with the relevant institutions.

For example, in 2006, the IUCN *World Initiative on Sustainable Pastoralism* (WISP) launched a study to demonstrate that policies, which support mobility and traditional governance, can deliver environmental benefits through enhanced pastoralism. To promote adoption of such policies a side event at a meeting of the United Nations Convention to Combat Desertification was convened in August 2007, bringing representatives of pastoralists from six countries to present their own experiences on the subject. However, it was observed that the event on its own would not be very influential and other opportunities were sought to achieve greater impact.

It was determined that a senior UN staff member was scheduled to make a presentation in the main conference. This member of staff accepted to refer to the findings of the WISP studies and make recommendations for their adoption by the UNCCD Committee on Science and Technology. Similarly, a number of National Government Focal Points were persuaded to endorse the recommendations in plenary discussions. This finally led to the chair of the CRIC recommending that pastoralism be adopted in the programme of work of the convention.[10]

This example indicates the importance of being alert and flexible to take advantage of opportunities as they arise to influence policy and dialogue. Some of these may be as simple and informal as lunch meetings or lobby talks that are crucial to complement planned policy interventions. However, the basis for implementing a policy-influencing programme should be a clear theory of change, which provides a sequence of activities that need to be implemented. These should take into account the factors that one is in control of and the external factors one may not be in control of but need to be monitored to determine how they can be influenced.

The lessons learnt from numerous policy influencing initiatives indicate that although elaborate theories of change were developed, many critical aspects were not followed up. Therefore, even if good results were often achieved, the processes are frequently not completed and as such the overall intended goals are either not achieved or are not observed or evaluated. One of the key causes of this is the short-term nature of projects under which the initiatives are implemented, which do not provide enough time for the processes to evolve and enable meaningful follow up. This is illustrated by the example in Box 6.4 of IGAD's efforts to influence dryland management programmes and policies across the region.

Some elements of the activities outlined in Box 6.4 were well implemented, and achieved the intended outcomes such as increased understanding and appreciation of the value of drylands amongst participating Members of Parliament

Box 6.4 *IUCN–IGAD partnership to influence policy*[11]

IUCN and IGAD have worked in partnership since 2005 to build awareness of the role conservation can play in sustainable development. Early interventions included research to build knowledge and create awareness about the opportunities and challenges, particularly in drylands. This research was used to raise awareness among politicians and public servants from the Ministries responsible for Finance and Environment in IGAD countries.

The studies were complemented by field tours of IGAD MPs to Garba Tula in northern Kenya where they interacted with communities to learn about their problems and recommendations. These meetings resulted in policy recommendations (see Chapters 2 and 4) which were captured on film and presented to a wider audience of Public Servants from IGAD States. This influenced the recommendations of those Public Servants and resulted in the development of action plans for implementation at national level. Monitoring of progress is ongoing, but recommendations have been captured in a follow-up initiative between IUCN and IGAD, which will facilitate further action on these recommendations.

(MPs), or greater capacity amongst communities to develop and articulate their policy recommendations to government. This was well demonstrated by the discussions between MPs after the tour and the policy recommendations that they agreed on (IUCN, 2008). These recommendations were also picked up by directors of economic planning and conservation at the 2010 IGAD conference in Entebbe and captured in the policy recommendations of this group (IUCN, 2010).

However, during a follow-up with MPs one year later, it became evident that a number of action points and recommendations had not been followed through by the participants. Even though it was evident that the directors and technocrats were aware of the issues and recommendations and in some cases had tried to follow up their action plans, it was obvious that they had needed further support to follow through the plans within their countries. This was a clear gap realized in the policy influencing process. The process had not planned to build on the genuine interest created during the regional meetings of the high level technical officers who were keen to account for their actions to fellow IGAD colleagues.

Interventions to follow through and build on the interest of the Government Staff at their national levels could have included support in undertaking analyzes and preparing briefs for presentation in budget meetings. When the Government Staff returned to their countries, many other pressing factors took priority, leading to the relegation of the action plans. If the strategy had identified in each country certain targets that would complement what had been achieved at the regional levels, there would have been a higher chance that the intended policy objectives would have been achieved.

Conclusion

Working through IGBs to influence policy is one approach that can be used to mainstream environment in development processes. However, it should be noted that the process takes time, is usually expensive and rarely directly results in actions at the national and local level. Therefore, the strategy to engage IGBs for policy influence needs to be evaluated well to ensure that it is the best approach to achieve the intended policy objective. The added value of working through an IGB needs to be made explicitly, compared to engaging directly at the local or national level. Values to consider include: achieving a wider reach of policy influence; promoting learning and harmonization of approaches; enhancing synergies to achieve national level impacts; addressing regional issues such as conflicts over resources that would be difficult to resolve nationally; championing of new ideas and concepts that other countries would be resistant to; and enhancing bargaining power.

To enhance success in achieving particular goals being pursued though IGBs, the strategy should consider any other key indirect values that are priority to the countries and can be achieved in addition to the primary policy influencing objective. These could include enhanced security, contribution to the implementation of national programmes, accessing financial and technical services among other things. This requires keeping abreast of each country's priorities, policies and policy-making structures, to identify those aspects that could be best delivered at the regional level. Being strategic will ensure that interventions remain relevant and effectively identify entry points: both in institutions and on issues.

In strategizing engagement through IGBs, their differences need to be recognized and those that are targeted for policy influence need to be carefully evaluated to understand their setup, operations and to determine if engaging them will enhance the target objective. If evaluated as suitable, then there is a need to determine the best approach to working with them and the costs and time required to engage them.

In conclusion, it is noted that IGBs can provide a potential avenue for influencing environmental policy, principally by creating a framework or an enabling environment that can catalyse action at the national or local levels. Engaging in policy through IGBs should ideally be linked to policy engagement at national or sub-national level in order to make the link between high-level instruments or frameworks and local level demands and pressures.

7

SHAPING POLICIES

Science–policy interface in natural resources management

Pablo Manzano Baena[1]

Introduction

The word *policy* can have different meanings and values to different people, depending on when it is used by a given person. For example, *public policy* may mean a decision, made by a publicly elected or designated body, to meet or satisfy public interest (Torjman, 2005). Kingdon (1984) considers it a set of processes that includes an agenda, the election of several alternatives in that agenda, an authoritative decision on the alternatives present, and the implementation of this chosen alternative. This is similar to the policy process sketched out in Chapter 1 in this volume. In this chapter we look at public policies for natural resources, those which guide and develop natural resources management, and how these policy decisions influence peoples' lives, livelihoods and the biophysical environment they live in, including biodiversity. Here, policy is therefore a public decision that is made to guide natural resources management in order to provide public goods and services, balancing at the same time various societal needs.

Society is supposed to be the ultimate beneficiary of scientific research (Leshner, 2002). But to generate this benefit, the results of research must be applied by policy-makers, practitioners, landowners, educators, other researchers or even private citizens. This is where the science–policy interface is situated, and its management can determine whether or not scientific knowledge is used effectively. The perspectives from where science and policy come are often very different, with policy needing quicker results, with greater certainty, and more focused on policy problems, than science usually provides (Wiseman, 2010). However, the impact of research on policy can be increased with an improved engagement of social actors, and in this way more value can be created more quickly, from scientific research for society (Guldin, Parrotta and Hellström, 2005).

The aim of natural resource policies is to guide the management of natural resources. In this respect programmes are made that have to be interpreted by policy decisions. The programmes, as well as the decisions, should be built on objective analysis of:

(i) the previous situation;
(ii) the problems to be solved; and
(iii) the methods to resolve these problems.

The foundations of science and the scientific method are based on procedures that theoretically guarantee the objectivity in the three mentioned steps. In this sense, policies that are built upon strong science-based evidence should provide accurate guidance both for social development and for the conservation of natural resources. The flow of information from scientists to policy-makers can occur both when policy-makers reach out to access information from academic and/or popular science publications, and when technical advice is pro-actively provided by scientists (Figure 7.1).

This theoretical framework is far from being confirmed in practice, and several questions arise. Are policies usually built on adequate scientific evidence? Is it always necessary to depend upon scientific evidence to develop a policy? Is it enough to just generate scientific evidence and expect it to be used to influence policy? What are the ingredients to ensure that scientific evidence is actually used to influence policy-makers' perspectives and perception? Does scientific evidence always have appeal to policy-makers? Does scientific evidence reach them? How can science be a respectable source or base to inform natural resources policy decisions?

Here we provide evidence for a positive link between science and policy in two parts. In the first, we examine examples that give evidence of science–policy interactions worldwide. In the second, we provide more localized experience, primarily from Eastern Africa, to illustrate where and how science has been effectively employed in policy processes.

Experiences worldwide

How science and technical advice are used to inform policy

Policies regarding natural resources should meet the expectations that society places on them, which are intimately related with people's livelihoods. They should be based upon accurate and relevant information, as policies define or at least set the aspirations on how societies interact with natural systems (Brown and McLeod, 1996). But the use of natural resources is constrained by the functioning of natural systems that is limited by physical laws. Therefore, natural resource management decisions are not only constrained by socioeconomic needs of the society, but

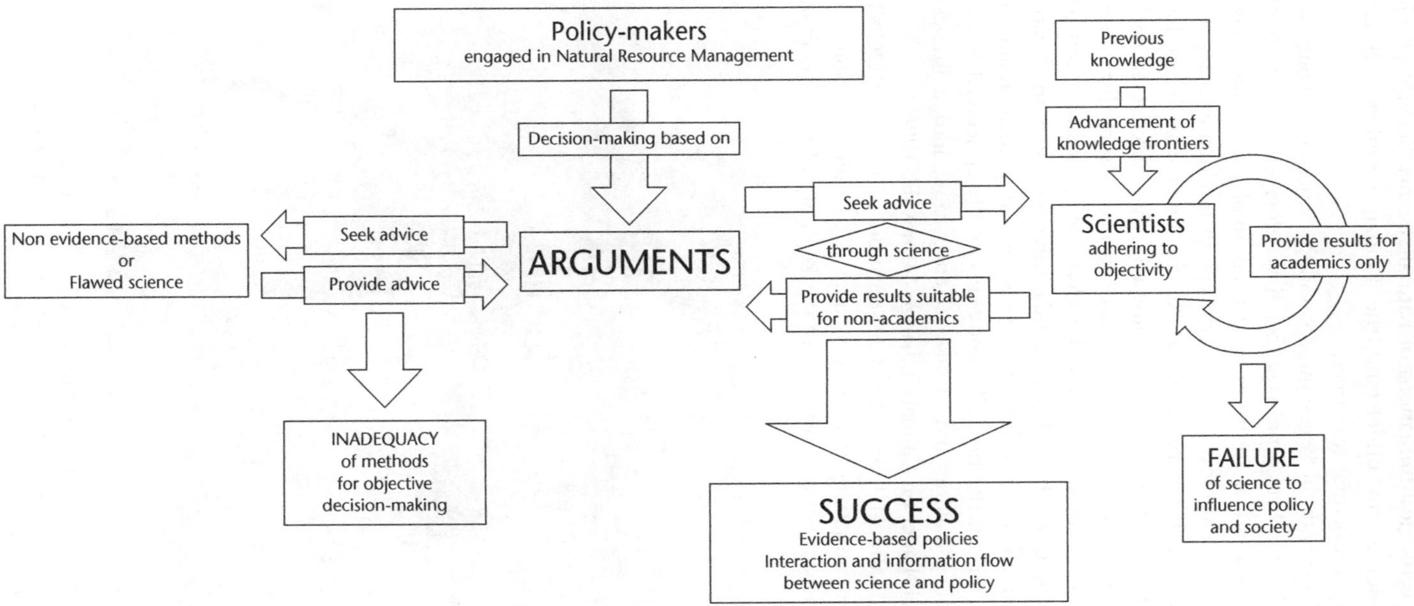

FIGURE 7.1 Positive and negative scenarios at the science–policy interface

also by the ecological limitations of natural systems. Science provides the basis and methodology to understand the human impulses and the associated interactions with natural resources.

The example with the largest impact among science-based policy-making comes from the climate change policies that have emerged as a result of the Kyoto agreement. As a global environmental issue involving a wide array of actors, it can provide examples for nearly every type of interaction between science and policy. However, the very origin of the climate change debate has its roots in pure scientific research. The greenhouse effect was discovered back in 1824 by Joseph Fourier and knowledge of it was deepened by such distinguished scientists as Tyndall and Arrhenius (Held and Soden, 2000), before it was perceived as a problem and entered the political agenda. Environmental organizations began to campaign for changes in fossil fuel policies basing their actions on scientific evidence. Later on, the first UN body specifically conceived to bridge the gap between policy-making and climate science came into existence: the Intergovernmental Panel on Climate Change (Agrawala, 1998).

The International Union for Conservation of Nature (IUCN) has several products and initiatives that have managed to influence policy through the scientific perspective, which is one of the key tasks of the organization. Its Red

FIGURE 7.2 IUCN's Red List of Threatened Species is an example of a scientific tool that effectively guides the conservation of activities of governments, NGOs and scientific institutions

List of Threatened Species, steered by the Species Survival Commission (SSC), has contributed to influence conservation policies worldwide by creating a systematic evaluation of species based on their conservation status and thus calling attention to those most endangered (Rodrigues *et al.*, 2006). The Red List is used to guide management of natural resources at multiple scales, from local Environmental Impact Assessments to international multilateral agreements such as CITES (Convention on International Trade in Endangered Species of Wild Fauna and Flora).

The World Commission on Protected Areas (WCPA) has done a similar work for the assessment of protected areas, establishing the World Database on Protected Areas jointly with the United Nations Environment Programme and the World Conservation Monitoring Centre (www.wdpa.org). Tools have been developed to assess the effectiveness of protected areas and their status, such as the Management Effectiveness Tracking Tool, among others (Leverington *et al.*, 2008).

The concept of Ecological Footprint has also been a driver of change in global policies. Envisioned in the early 1990s by William Rees and Mathis Wackernagel, it is an exercise on relating the appropriation of resources by humans with the total carrying capacity of the planet (Hoekstra, 2009). It has been used by for example the European Commission as a touchstone for relating resource use and carrying capacity, complementing it with other indicators (Best *et al.*, 2008).

More subtle policy influencing processes can also take place, steered at times by field practitioners. In Niger, pastoralists have been heavily repressed since colonial times, to the extent that they tended to hide their opinions in public forums and were more likely to state whatever they thought officials expected from them. Scientific arguments on the soundness of mobile pastoralism have effectively triggered the self-esteem of pastoralists and have allowed them to dare to present their personal views (Sommerhalter, 2008:171). This way, they are able to be taken into account by policy-makers and have contributed to the development of pro-pastoralist policies.

Interestingly, in all examples provided we find some intermediary actors that catalyse the knowledge from scientists to policy-makers. Slinger and Vreugdenhil (2007) found that the existence of appropriate scientific knowledge alone is not adequate to influence policy or to ensure transition to improved management of natural resources. They suggest that, to influence policy, scientific knowledge needs to be supported by cohesive scientific groups and championing individuals to build bridges and transfer knowledge. It is also necessary to have strong institutionalized communication of scientific information for achieving evidence-informed policy-making in developing countries (Jones *et al.*, 2009). An understanding of the societal institutional context is also complementary in the knowledge transfer process. It is also important to acknowledge the barriers to research in affecting policy so as to simplify the process of building strategies around the dissemination of research so that it becomes more accessible to policy-makers for easier adoption (Walt, 1994).

How policy and practice influence science

Even if the results of research inform policy processes, research itself needs funds to be conducted and therefore depends on political will for its financing, either from the public sector or from some private actors. This is why it is important that the research agenda addresses issues relevant to society as presented by the agendas of politicians or decision-makers, and sometimes even to challenge them. Therefore, if scientists want to have an impact on policy and generate interest in their research, they must carefully assess what research and what research question is relevant to the most pressing policy issues that relate to natural resources management (Guldin *et al.*, 2005).

The world is suffering a widespread crisis of biodiversity loss that boosted the establishment of the Convention on Biological Diversity at the 1992 Earth Summit. However, nearly 20 years after that, the loss of biodiversity has not been halted. The mechanisms regulating biodiversity are often complex, as is the identification of critical thresholds. Political action to preserve biodiversity requires clear action targets that, until recently, science has failed to provide. This clear demand by policy, however, has driven the identification of targets categorized by their urgency and their long-term need (Mace *et al.*, 2010).

Returning to the climate change issue, the amount of resources spent in its research so far is an example of how a very big portion of the available research resources is directed by global political interests. Understanding the causes and impact of global climate change is a complex matter; however, research investments have enabled this subject to be at the top of global political decision-making. This is because a good number of climate change scientists and policy-makers alike are convinced that the ramifications of climate change are far-reaching. The directions in which to orient the research that is carried out, and thus the way global climate change discussion is influenced, are nevertheless polarized. Industrialized countries, which have significantly contributed to the anthropogenic causes of climate change, are also the ones that have capacity to carry out and finance research and therefore to shape international policy in climate change. Discussions on the issue have consequently been dominated by mitigation debate and how to cut down on greenhouse gas emissions. Conversely, in less industrialized countries climate change research has been a low political priority and has received very little local funding (Kandlikar and Ambuj Sagar, 1997). Only recently has investment increased, with a focus on adaptation techniques and on understanding the impacts of land use change towards confronting climate change, especially in the world of development. This is especially true in areas where the impacts can be deep but resources for research are scarce, such as in drylands.

Moreover, scientific research can get away from its core philosophy if the funder-driven pressure becomes too high. The oil industry initially aligned strongly along *climate sceptic* lines, and even if there has been a major change in its positions (Kolk and Levy, 2001) the Republican Party of the US continued to support *climate sceptic* views at the turn of the century (McCright and Dunlap, 2003). This support even translated into direct sponsoring of many *climate sceptic* studies by conservative think tanks that would otherwise not have been funded, and that reinforced *climate sceptic* and anti-environmentalist positions worldwide due to their scientific character (Jacques, Dunlap and Freeman, 2008). This adds to the complexity of changing views on climate change, even among a non-academic but interested audience, given the long-term effects that are not perceivable in the short-term. This usually causes the audience to accept scientific arguments that reinforce the thesis they supported previously, but to reject the ones that belong to the thesis they disagree with (Lorenzoni and Hulme, 2009).

An older example is the Lysenkoism doctrine which, based on ideological precepts, rejected the role of genes in inheritance. Lysenkoism was developed during the rule of Stalin and applied to agriculture, and its disastrous consequences lasted for decades in the Soviet Union and China (Li, 1987). However, post-war Eastern German scientists were generally more sceptical of mixing ideology and science and as a result the doctrine never rooted there (Hossfeld and Olsson, 2002).

Scientific research can also be driven in a positive direction by political agendas, such as in the case of pastoralism. The food crises in the Sahel during the 1970s and 1980s triggered research on the area to analyze their causes. Early on, the conclusions of the first studies pointed towards pastoralism being a highly productive and sustainable livelihood that was affected by the closing of international borders in the area (Sinclair and Fryxell, 1985). The accumulated scientific corpus since then has ultimately led to a feedback for policy influencing, materialized in the endorsement of the African Union Policy Framework for Pastoralists (Schlee, in press). Despite this outcome, it should be noted that the case of pastoralism, particularly in sub-Saharan Africa, gives a strong illustration of a politically-motivated agenda that has skewed research towards alternative land uses that are less economically, socially and environmentally sustainable (Davies *et al.*, 2010).

Challenges and gaps in using science to inform policy

The extent to which research influences natural resource management and conservation policy is widely debated. Researchers often argue that good scientific results and knowledge are neglected by policy-makers, who often find it challenging to understand complex, often non-linear relationships, and thresholds found in nature. The percentage of area to be put under conservation schemes for guaranteeing species survival or delivery of ecosystem services, for example, is by no means a constant figure across species, ecosystems or regions, but policy-makers tend to simplify the problem by putting 10 per cent of a region or country area as a target for conservation, thus greatly understating the real need for conservation areas (Svancara *et al.*, 2005). Conversely, policy-makers complain that decision-making in policy development is hindered by conflicting findings, confusions and disagreement among the scientists (Walt, 1994).

Policy-makers may not always be inclined to access scientific knowledge to support their policies, sometimes preferring to pursue an agenda that is influenced by other actors. However, when policy-makers are inclined to seek out science, they may find the relevant information lacking, inaccessible, difficult to interpret, or not coming from a legitimate source. For example, research may be carried out by the World Bank or a foreign research institution and may be rejected as being irrelevant to the local context (Carden, 2009).

This debate is obviously influenced by different approaches towards knowledge. Policy-makers "use" knowledge, and they usually need certainties so that they can guarantee the taxpayer that the decision they are taking is the correct one. On the other hand, the scientist ideally performs research out of curiosity, and science is surrounded by the uncertainty of the scientific method (van den Hove, 2007), which states that any hypothesis, theory or law is always subjected to validation irrespective of how long it has been considered valid. There will always be an element of uncertainty that decision-makers need to deal with (Wardekker *et al.*, 2008). Ecosystems are especially complex (in the sense that they are unpredictable and extremely variable), so the area of natural resource management is particularly prone to this conflict. The preference for certainties may explain the preference of policy-makers for technical staff (e.g. an agronomist) compared with scientific researchers (e.g. an ecologist).

Science needs to be communicated and understood

Communication is a key task to bridge the gap between subjective perceptions and objective knowledge (van den Hove, 2007). Ancient negative cultural connotations of the dune desert for Han Chinese, for example, justify the settlement policies for pastoralists in Inner Mongolia or Xinjiang. In the Republic of Mongolia or Kazakhstan in contrast the positive cultural connotations for Kazakhs or Mongols lead to greater acceptance of and support for livestock mobility. These cultural barriers also exist in many East African countries and make it very difficult to accept the widespread scientific evidence in favour of mobile pastoralism (Zukosky, 2007). Popular science has a great capacity to alter those views, but it requires scientists to engage in activities (writing for popular science journals, or spending time in collaboration with journalists) for which they are often not paid.

Another important factor that sometimes fuels the conflict between scientific and political views is the scale of the decisions, geographical as well as temporal. Politicians are, as mentioned above, strongly results-oriented, and they usually require quick results to maintain their approval ratings, i.e. not making unpopular decisions. This applies to democratic regimes, subjected to 3–5 year long political mandates (Wiseman, 2010), but also in non-democratic regimes interested in keeping social peace. They therefore may pursue short-term benefits, even if they translate into long-term losses. A good example for problems with geographical scale is the policy for encroachment of agriculture in the dry-season pastures in Eastern Africa (Aboud *et al.*, 2012). While in the short-term the cultivation in the wetlands may locally mean a higher profit per hectare, in the long-term it means a disruption of the pastoralist system. The herds lose their dry-season feeding grounds, the animals have to stay year-round in the dryer areas and in a few years their numbers drop as the carrying capacity of the ecosystem sinks dramatically, thus lowering the income per hectare if the entire ecosystem is considered (Davies and Hatfield, 2008).

More frequent is the incapacity to interpret natural resource management at a relevant temporal scale. For that, many of the most dramatic examples come from fisheries management (Worm *et al.*, 2009; TEEB, 2010). The collapse of the Californian sardine fishery, directly followed by the collapse of the Peruvian anchoveta industry, is one of them. Fishing fleet numbers, and thus maximum yields, are adjusted to years where the conditions are particularly good, but in bad years the reduction of the fleet gets very challenging (Ludwig *et al.*, 1993). Complexity is added by the fact that determining the maximum sustainable yield of a fishery does not seem possible, as Ludwig *et al.* write, "without previously overexploiting" it. Additionally, the pressures from the fishing industry to influence research are frequent.

Challenges of economic valuation

The economic valuation of the ecosystems is a discipline that shows how complex the challenges in the science–policy relationship can be. These studies were originally developed in order to better inform policy decisions on "incremental" or "marginal" values of ecosystem services as well as to orient business on risks and opportunities for them. Understanding the total value of ecosystems is imperative for decision and policy-making. Scientific research has shown that it is important to maintain the health and integrity of ecosystems; and to balance investment in order to optimize and sustain its economic output. But many policy-makers, investors and even ordinary people believe that maximization of productivity through investments that heavily intensifies ecosystem use from their traditional uses are the best ways to improve the economies of their inhabitants and to take them out of poverty. This misconception is perpetuated because most policy-makers have the opinion that traditional uses are not productive enough and that traditional users are chronically poor (Costanza *et al.*, 1997; TEEB, 2010).

Research, has shown, however, that the traditional methods are productive and sustainable in supporting livelihoods, especially in non-monetary terms. Estuarine areas in the proximity of Shanghai, for example, have shown a high providence of ecosystem services during millennia that are fundamental for their traditional users, and that could also be fundamental for the city in terms

FIGURE 7.3 A desiccated maize crop in Kaabong, Uganda: farmers were persuaded by extension agents to adopt a new crop variety that exposed them to unacceptable levels of climate risk (Nangiro, 2005)

of long-term protection against high tidal surges (Zhao *et al.*, 2004). But the troubles of performing an effective valuation of the ecosystem services there have caused the city government to develop some land use changes that have reduced the ecosystem service provisions of the areas, even against its environmentally friendly purposes. Similarly, there were plans to drain the Nakivubo Swamp in Kampala, Uganda, for real estate development, without taking into account its high value for the purification of the wastewaters spilled by the city before they entered Lake Victoria. The plans have been finally discarded following the economic valuation of the services provided by the whole swamp at 1–1.75 million US$ per year, with further restoration planned that will add value to the ecosystem (TEEB, 2010).

The undervaluation of traditional practices has its roots, again, in complex realities. Policy developers and economic planners do not adequately consult traditional users, and scientists have not appropriately presented the findings in communicable messages to policy-makers. But elaborating a Total Economic Valuation of an ecosystem demands a complex conceptual framework that is not easy to understand by many actors. Policy-makers tend to value economical systems by their pure monetary value, and even users of the ecosystem services are not aware of them if they don't pay for them (TEEB, 2010). Barter-like non-monetary transactions are frequently missed, and they often represent a very big portion of rural economies (Davies and Hatfield, 2008). Other intangible values, much more complex to valuate but often more important, are also provided by traditional uses. Unsustainable use of rainforests can provide higher private revenue, but the social and overall benefits are much higher when sustainable uses are implemented (Turner *et al.*, 2003).

Understanding and using traditional knowledge

Knowledge is very important in decision and policy-making regardless of the type and the purpose of the knowledge: i.e. whether it is produced for economic, social or political development. In the development discourses knowledge is typically viewed in two paradigms: indigenous knowledge or scientific knowledge (see Chapter 2 for a more comprehensive discussion of this subject). According to Grenier (1998), indigenous knowledge is the unique, traditional, local knowledge existing within and developed around the specific conditions of women and men indigenous to a particular geographic area. This is also referred to as "traditional" or "local" knowledge.

The term refers to the large body of knowledge and skills that has been developed outside the formal educational system, usually across generations and through tried and tested procedures, and without necessarily implying an understanding of the underlying mechanisms for the application of this knowledge (Berkes *et al.*, 2000). On the other hand, science is the system of knowledge, which relies on certain laws that have been established through the application of the scientific method to understand phenomena in the world that surrounds us. The process of scientific method begins with an observation followed by a prediction or hypothesis, which is then tested. Depending on the test results, the hypothesis can become accepted by the world scientific community (The Living Knowledge Project, 2008). Scientific knowledge has a broader following than indigenous knowledge, because of the argument that scientific investigation is prone to analytic reasoning and that it is universally applicable (Agrawal, 1995; Milani, 2009; Guchteneire *et al.*, 2010).

However, the tendency of policy-makers to favour technicians (in contrast with scientists, as mentioned in the example of the agronomist and the ecologist above) when looking for advice distorts this vision. This knowledge advice can be distorted by conflicts of interest or by the manner in which it is generated (Agrawal, 1995). This also includes complying with deadlines or with request for certainties where it is not possible to get them. Additionally, the findings of traditional and scientific knowledge can be largely complementary. Specific strategies for protecting, systematizing and disseminating knowledge will benefit different groups of people in different ways (Agrawal, 1995).

In some extreme cases, the application of technologies that have disregarded the pre-existing traditional knowledge are considered to be the cause of many of the problems that we now face in natural resources management (Mander, 1991). Therefore, indigenous knowledge is increasingly viewed as complimentary to conventional scientific knowledge for informing policies and solving problems in natural resources management and sustainable development (McGregor, 2004; Guchteneire *et al.*, 2010).

Co-management of natural resources provides a practical framework for linking science and local knowledge, where local knowledge and receptivity of policy-makers are as important as bridging organizations (Berkes, 2009). Caution should be exerted, however, in relying exclusively on indigenous/traditional knowledge. Given the reduced costs of documenting it compared with proper scientific research, it can be tempting to base decisions exclusively on it. Limitations apply to traditional knowledge, especially in changing environments. One of the rules of science is "discard hoary old shibboleths when evidence shows they are misguided" (Weiss, 2009). Nevertheless it is highly convenient to understand traditional knowledge through science, and learn from it.

Constraints for research–policy interaction: a focus on Eastern and Southern Africa

Most scientific research institutions in Eastern and Southern Africa are public institutions. These include universities, specialized research institutes and conservation or development organizations that conduct research. In most cases the financing for the natural resources research is dependent upon international funding through multilateral or bilateral aid. For this reason the research agenda is usually responsive to the interests of the funding institutions. The danger of the need for short-term results as a major driver for the research agenda is consequently present.

Similarly, policy development is the responsibility of public institutions. The process is usually led by technical advisors but approved and endorsed at a political level. In recent years, the processes to develop natural resources policies has been made increasingly participatory and thus has endeavoured to involve multiple stakeholders, including the Civil Society. The policies are thus documented for national consistency and generate common understanding of goals and pathways to natural resources management. It should be noted that the implementation of the policies is in most cases done through projects that are also supported by international donors.

While the research results may be intended to inform policy, it is possible that the results attained may be irrelevant to the policy objective of the country, because they are designed to meet the financier's objectives. Conflict of interest can also come from the same national governments, which can promote inappropriate concepts in order to get resources. An example is the common assumption of the advancing of the Sahara Desert in West Africa and widespread government support to halt it, despite empirical evidence demonstrating that on the contrary the desert has been on the retreat for the past two decades (Mortimore, 2009).

Policy-oriented research can be driven by international priorities and paradigms that are not necessarily relevant to the local context. In African rangelands, for example, research and policy has been strongly influenced by a northern understanding of soil mineralization by soil microbes, which is a common process in temperate latitudes. However in the local context, such processes are driven by rumen microbes and therefore soil mineralization depends on herbivore presence and action, with major implications for stocking and management practices (Ruess and McNaughton, 1987). As a result, research and policy in African rangelands has frequently driven inappropriate practice that has contributed to degradation and poverty (Davies *et al.*, 2010). Research typically centres on "easy to spot" and "easy to sell" targets, so that the research is cheap and can be visible to donors (Manzano *et al.*, 2010). Biodiversity studies often focus on wildlife but do not study underlying ecology, for example, the invasive plants that can affect the trophic chain and have dramatic consequences for the herbivores that are valuable for the tourist industry (IUCN, 2010).

A further challenge is that there usually is no permanent system for researchers to inform policy-makers of their findings in such a way as to influence the policy development process or the policy outcome. Research is infrequently commissioned by policy-makers in the region, and research institutions may not be recognized by policy-makers. The adoption of science into policy discourse is further frustrated by challenges over timing. Policy-makers may ignore researchers until science is urgently needed, by which time it is too late to conduct the requisite research. Even when science is available in the right format and on time, policy-makers do not always have the capacity to interpret results and use them to effectively adjust policy.

Positive examples of research–policy interaction in Eastern and Southern Africa

Research can play multiple roles in influencing policy and practice at different levels of government, and other chapters in this book have illustrated ways that non-research actors can wield science to support their advocacy or lobbying in favour of particular policy outcomes. Chapter 9 returns to the discussion of the formation of appropriate alliances, not simply of like-minded institutions but deliberate alliances of different institutions with different skills and roles. These alliances would include research partners and advocacy partners, who have different credibility and legitimacy in the eyes of the target audiences, and have different avenues for policy influence.

Research can play a role in refining the arguments that policy-makers wield, in introducing new terminologies or new concepts that improve the basis of policy dialogue, or introducing or cultivating new champions or experts to the debate. Research can bring new ideas into policy debate and broaden the overall scope of a policy debate, and if conducted and communicated effectively can expand the intellectual framework surrounding policy-making, and open up avenues for identifying innovative solutions where policy-making is confounded by seemingly dichotomous choices. Finally, research can play a role in the way policy decisions are made as much as in influencing the actual content of policy, lending legitimacy and improving buy-in from a wider range of actors (Carden, 2009; Swanson and Bhadwal, 2009).

Policy-making in most countries develops through small steps rather than radical change, which generates a strong preference for research that supports existing paradigms and positions as opposed to truly innovative findings. Policy-makers are generally less interested in revolutionary research and therefore radical changes require much more sophisticated approaches to influence the policy discourse. Policy-makers (and society at large) are likely to be resistant to research that undermines long-held assumptions or beliefs, and therefore researchers should take a longer-term strategy of building capacity and sensitizing decision-makers (Carden, 2009).

Based on these insights, Carden (ibid., p. 23) proposes five recognizable categories of research and policy interaction:

1. Clear government demand.
2. Government interest in research, but leadership absent.
3. Government interest in research, but a capacity shortfall.
4. A new or emerging issue activates research, but leaves policy-makers uninterested.
5. Government treats research with disinterest, or hostility.

The most suitable approach to introducing research to the policy discourse will change according to these categories, and examples from other chapters in this book illustrate some of the ways that researchers can strategize influencing policy, particularly where government is disinterested or hostile. The role of Civil Society and political leaders in popularizing new ideas, for example, can be crucial in gaining traction for major policy reform, or for demonstrating success on the ground that can reinforce policy debates at a national level.

Clear government demand

Clear examples of government-commissioned research to influence policy are not as widespread as might be expected, highlighting the fact that research is not necessarily highly valued in policy development. However, where government champions the development of a particular policy it may open avenues for research that can feed into the process, either directly or indirectly. Often the challenge in this case is mobilizing research in the short time that may be available to respond effectively. Indeed, the time required to conduct rigorous research may be a factor in the low demand for research from governments that want to push a policy through in a short space of time.

Uganda's Wetlands Policy provides an example in which the development of policy stimulated or drew on relevant supporting research (Chapter 5). Wetland resources in Uganda have traditionally been used for construction, craftwork, furniture, and as hunting and fishing areas. Seasonal wetlands and the margins of permanent wetlands are used for grazing cattle, growing crops and as a source for domestic water. In addition, they are an important habitat for wildlife. Protecting these valuable habitats and economic resources was a priority of the incoming National Resistance Movement Government in 1986. In September 1986, Government issued administrative guidelines to reduce damage to wetlands, including a ban on large-scale drainage schemes "imposed until such time that a more elaborate, scientifically proven and socially harmonious policy was put in place" (Republic of Uganda, 1995). Government and researchers collaborated in creating a national Wetlands Inventory and associated research that provided the basis for environmentally sound management and rational utilization of the wetlands resources. This research guided the subsequent development of the National Wetlands Policy and adoption of wetlands later still in a number of national laws (Mafabi *et al.*, 2005).

Government has interest in research but lacks leadership

Where an issue is of known importance, but no individual or department has taken the lead in championing the debate, there are a number of ways to stimulate leadership. Chapter 2 (Sophie Kutegeka and Guyo Roba) in this book highlights how communities have championed issues locally and have networked to raise their concerns at higher levels of government. It is possible to stimulate leadership within government, both political leaders and technical advisors, through citizen-based advocacy and support to bring decision-makers and communities together.

Chapter 4 in this book (Ben Wandago) presents a number of ways that elected leaders can influence policy, particularly by setting policy agenda's, and this is an important way to influence leadership within government departments. Chapter 6 (Barbara Nakangu Bugembe) documents experiences of working with Inter-Governmental Bodies to develop region-wide policies and positions, which provide another approach to stimulating debate on a new subject and promoting leadership within a particular government institution, or promoting the work of champions within those institutions.

The Indian Ocean Fishery Commission provides an example of stimulating political leadership through an inter-governmental institution, and using that institution to commission and champion research. Fisheries in the Indian Ocean are one of the main resource activities for some island countries such as Comoros, Mauritius or, especially, the Seychelles (Robinson and Shroff, 2004). In the latter, most of the population rely on traditional fisheries for their food security, even if there is a well-developed tuna fishing industry. Current governmental efforts point to the development of a semi-industrial fleet that can share the vast resources of the surrounding waters with the foreign companies present, mostly European (purse seine) or East Asian (long line) (Alexis and Chang-Sam, 2006).

The region has a long history of overexploitation of fisheries, but has been subjected to the control of the Indian Ocean Fishery Commission since the 1960s. The tuna fisheries have been specifically controlled by the India Ocean Tuna Commission (IOTC) since 1989, which was created for this purpose (Kambona and Marashi, 1996). The IOTC is an FAO-created commission, which includes amongst its members government agencies such as the Seychelles Fishing Authority (Alexis and Chang-Sam, 2006), and which plays the role of controlling the size of the fleet in the area. Detailed studies such as the Indian Ocean Tuna Tagging Programme (Haillier, 2008) have been conducted to effectively monitor the tuna populations in the area.

The application of control measures seems to have yielded positive results in term of sustainability of tuna captures during the last years (Polacheck, 2006), even if some criticism has been raised (WWF, 2009). So far, the Indian Ocean tuna fisheries seem to have had a far more sustainable use than other fisheries (Robinson and Shroff, 2004), and some other fisheries in the world are following its trend, highlighting the role of governance and scientific evaluation (Worm *et al.*, 2009).

Government has interest in research but lacks capacity

Lack of capacity (whether a lack of technical skills or inadequate human resources) in government may render government staff unable to commission relevant research, or to identify relevant research that already exists, or to interpret the research that is put in front of them. Obviously, particularly in the last of these examples, researchers can do a great deal to make their research more comprehensible. However, where government agencies lack capacity to take the lead on policy-oriented research this may generate discomfort and a sense of disempowerment. Researchers can also work on this to build relationships, and to genuinely consult and engage decision-makers in research processes. Chapter 8 in this book (Hesselink and Zeilder) provides a rich array of communication approaches that can be used, according to the policy-research relationship, to effectively build capacity in government.

An example of building government capacity to use and commission research is the inclusion of Environmental Flow Assessments (EFAs) in water policies, for example, in Tanzania (Hirji and Davis, 2009). When funding high-impact water infrastructures, such as dams, the World Bank accounted for upstream impacts, but it was not until the mid-1990s that downstream impacts were taken into account. The increase in methods to estimate environmental flow requirements and the increased knowledge on the ecological response to different flow regimes led the Bank to include EFAs in its Environmental Impact Assessments and influence developing countries in including environmental flows into government water policies. This is done in collaboration with different environment related international NGOs and agencies.

In Tanzania, following major crises in water resource uses in several basins (Great Ruaha, Pangani), scientific studies have helped understand the collapse of fisheries and the crises in hydropower generation, mainly caused by the aim to get short-term profits. As a result, a National Water Policy was approved in 2002 to regulate water for basic human needs and environmental uses. The country has created several river and lake basin offices that are meant to control water resources in every basin. The influence of scientific knowledge on the policy can be seen in quotes such as: "water for the environment shall be determined on the best scientific information available considering both the temporal and spatial water requirements to maintain the health and viability of riverine and estuary ecosystems" (Ministry of Water and Livestock Development 2002, p. 15), or "water resources assessment will be done on the basis of sound scientific and technical information and understanding" (ibid., p. 21). Importantly, one of the lessons of the Tanzanian experience in the cited study is that EFAs can be institutionalized prior to having supporting legislation, if there is enough will and technical support from the authorities.

An emerging issue activates research, but policy-makers are uninterested

In this context the challenge is one of popularizing or raising awareness of an issue that is considered to be of importance, but which leaves decision-makers, as well as the general public, indifferent. Climate change is probably the biggest environmental policy debate on the planet at this moment, yet not so long ago it was not in the public consciousness and attracted only minimal policy interest. Sometimes an emerging issue stirs immediate interest, as in the case of Genetically Modified Organisms (GMOs) which quickly led to the development of a regulatory policy in Kenya[2] and Uganda[3] and official guidelines on GMOs in Tanzania.[4]

A more localized example is the Kibale and Semliki Conservation and Development Project (KSCDP) in Uganda. The Kibale and Semliki are two National Parks located in the Albertine Rift Valley in western Uganda. These conserved areas, located in forests, faced some challenges related to the lack of knowledge about the functioning of their ecosystems. Local authorities partnered with IUCN and the local communities, funded by the Norwegian and the Dutch

Aid, in order to analyze the problems in a scientific way and then apply the results for the park management (Chhetri *et al.*, 2004).

One of the problems analyzed was the human-wildlife conflict in the park edges, given the absence of a buffer area around the parks. For Kibale National Park, several deterrent options were examined on the field following scientific protocols, and a holistic approach was applied in their evaluation. Although the building of trenches was the most effective alternative in the short-term, it was discarded because it displaced the problem elsewhere and it cut the migration routes of some of the problematic animals that constituted an important resource for the park, i.e. elephants. Finally, the plantation of buffer crops (coffee, soybean and cocoa) was chosen as the best and most cost-effective alternative and it was implemented in the area (Chhetri *et al.*, 2004).

Collaborative Resource Management (CRM) was also implemented in the project area as a means of linking government decision-makers directly to the research on the ground. Local populations had been prevented from using their traditional gathering grounds following the declaration of Kibale Park, and their attitude towards the protected area was thus very negative. Four pilot CRM sites were established so that the local community could recover the use and the voice on some resources from the park that could be harvested sustainably. The pilot projects were conducted in several phases: community sensitization, user identification and focused discussions, negotiation of terms and conditions and committee formation, sharing the draft agreement with other stakeholders, and signing the agreement. Monitoring and evaluation of the process concluded that CRM is more successful if implemented with other income-diversifying initiatives, that it needs human and financial resources to work in the long-term, and that it has managed to reduce the illegal activities by involving locals in the conservation of resources (Chhetri *et al.*, 2004).

The strengths of this initiative were the involvement of the Government (in the form of the Uganda Wildlife Authority, at a national scale, and the parks' authorities, at a local scale) as well as of the local population that participated in the whole process. The experience has been successful enough to replicate it in other Ugandan protected areas such as Mt. Elgon National Park (E.G.C. Barrow, personal communication; Chhetri *et al.*, 2004).

Government treats research with disinterest or hostility

As discussed earlier in this section, there are times when government may actively discourage research, or may not want to be told that the existing policy framework and agenda is inadequate, preferring to make small incremental rather than large fundamental changes (Carden, 2009). Sometimes what we refer to as government may in fact be a particular ministry, and support from within government may be found from a different ministry – this situation may arise frequently where agricultural and environmental policies are at odds.

In such situations, research findings may recommend that a policy agenda is modified, which requires political support often at the highest level of government. Under such circumstances it may be necessary for civil society, elected representatives and citizens in general to make the push for a shift in the policy agenda. However, researchers can still play a role in these processes, providing supporting evidence and raising awareness amongst activists of new arguments and policy options.

The recent African Union Framework Policy on pastoralism provides an example of a policy that has been developed through an Intergovernmental agency, has ensured both rigorous scientific input and effective community consultation, and has developed a policy agenda that is at odds with the national agendas of some member states (Schlee, in press). As mentioned earlier, many policy-makers have perceived mobile pastoralism as a primitive and problematic livelihood that should be eradicated, despite proven economic, social and environmental sustainability in the areas where it is traditionally practiced. The corpus of knowledge generated during the last 30 years has made it possible to re-evaluate pastoralism as a sustainable livelihood, and research has been widely publicized in the last decade by organizations such as IIED, SOS Sahel, the World Initiative for Sustainable Pastoralism, and the IUCN Commission on Environmental, Economic and Social Policy. The outcome is that African countries have adopted this resolution, which considers pastoralism as a development strategy and not as something that should be abandoned in order to achieve development. Although the policy is not binding, it nevertheless provides a framework for a paradigm change. Already existing positive policy examples from the last decade are included in the document, such as the mobile delivery of education in Chad or Kenya, pastoral laws in Mali and Mauritania that protect pastoral mobility, or the agreement in ECOWAS for transboundary herd movements. The approval of such documents illustrates a general and widespread change of perception amongst the diverse collection of actors we call "policy-makers" and highlights the long-term results of systematic research and communication.

Conclusion

It is a general assumption among scientists or science-oriented practitioners that science should inform policy. This is because it is considered that science is objective and methodical, and that the rigour of scientific methodology produces facts-based evidence. It is considered that policies developed from such evidence will definitely be the correct ones, because they will tackle the very causes of natural resource management problems that are addressed.

Unfortunately, research in natural resources generally takes a long time to finalize, more so when it is adaptive research. On the other hand, policies are developed to solve immediate problems for the long-term. In most cases policies are developed to appease political lobbies. In such circumstances science is frequently disregarded and not used to inform policy, especially when science results conflicts with political interests of policy-makers.

Another major challenge is the poor ability of scientists to communicate their findings beyond the realm of science. In most cases scientific forums are used as venues of communication between scientists. Policy-makers receive scientific documents that are almost invariably difficult to comprehend and inadequately packaged for policy science dialogue. Additionally, popularization of scientific results is not valued in the academic curriculum.

Positive examples, however, reveal that science can reach policy-makers and influence their decisions, and can boost their interest even to the point of building up specific platforms for the policy-science interface in order to resolve key challenges. A key factor is the format in which scientific results are formatted for the broader public. The interest of the broader public opinion is key to awaken political interest in science or, alternatively, the interest of policy-makers to bring scientists to the table in order to contribute with advice and solutions. Even if democracy and freedom of speech make it easier for some stakeholders to press governments with objective scientific data to achieve fair decisions, there are examples of non-democratic governments that also accept advice on guidance based on scientific data. This shows the potential of science to steer policies positively.

Thus, while science is valued as a strong basis for policy formulation, how science is carried out and communicated is of utmost importance in sending the right message to policy-makers. Communication of science requires highly innovative thinking, and should encompass the formation of alliances, not with other scientists, but with non-scientists who engage with policy-making in completely different ways. However, scientists can also do a lot more to engage decision-makers in research from its inception, giving them a greater understanding of and influence over the research agenda. There is a balance here between generating ownership of science and avoiding political capture of science. The integrity of scientists and scientific methodology is paramount, as it is always in danger of being influenced by funding sources or by diverse lobbies.

8

BRINGING CONSERVATION SCIENCE TO LIFE

The role of communication to support policy influencing processes in Africa

Frits Hesselink and Juliane Zeidler

Introduction

This chapter is based not so much on literature research as the other chapters of this book, but it is based on years of experience of the authors as consultants and on examples provided by their networks[1] and by some of the other authors[2] of this publication. The chapter explores how communication can most effectively support the process of policy influencing. It will demonstrate the need to 'bring to life' the scientific research and evidence based knowledge of conservation, before it can influence policy. The chapter will demonstrate that effective communication to influence policy is much more than policy briefs, reports, glossy brochures or other communication tools. It is about setting the stage, creating the mood, crafting the right language and images, finding the right time and right means to deliver the messages. And most importantly clearly state what specific actions are expected from the different audiences.

The chapter first looks at eight common errors made in communicating to influence policy. With awareness about the power of communication, the proper skills and strategies, these errors can be overcome. For the conservation community the key is to understand the basics of communicating to an audience that is not passionate about conservation, but has the power to take decisions for the better or worse. The chapter then provides examples from the African region of communication to various key audiences in the policy influencing process.

Eight common errors in communicating to influence policy

It has been our observation that conservationists often complain that many policies do not adequately reflect the science of conservation biology and the evidence based knowledge from practical conservation projects. They further complain that policy-makers do not listen enough, and sometimes not at all, to the science. Here we try to analyze to what extent communicating conservation science failed to influence policy and may continue to do so.

From our experience as consultants we noticed the following most common errors:

1. Letting the facts, figures and other evidence speak for themselves.
2. Using communication as an add on and not integrating it in the project.
3. Not being aware of the principles of systemic change.
4. Forgetting that influencing policy means influencing people.
5. Using messages that do not stick.
6. Applying wrong communication approaches or wrong expertise.
7. Forgetting to develop a strategy.
8. Sticking to old-fashioned prejudices regarding spin, style and PR.

Letting the facts, figures and other evidence speak for themselves

To influence policy conservationists often tend to supply their audience with as much information as possible. The more information, facts and evidence supplied, the easier policy-makers will be convinced, they seem to reason. It seems the conservation community assumes that everyone in this world is passionate about conservation, or at least will be once they are informed. What they do not realize is that policy-makers often do not ask for their expert information. They are busy people with many other priorities and concerns. When at a first glance the information offered is too scientific, does not connect with their immediate priorities or is not at all appealing to read, they will consciously or subconsciously decide that the information offered is just not relevant for them at this moment and the report ends up on the shelf. This happens all the time in offices of decision-makers all over the world.

Writing policy briefs and reports might be important but is not enough. When we analyze the situation in more detail we find that it is an assumption that the briefs will be read. If they are read it is an assumption that the messages of the brief or report will be understood. If they are understood it is equally an assumption that they are agreed with. And if the audience agrees with the messages, it is finally an assumption to think that they automatically will be acted upon. Effective communication deals with these assumptions in advance.

To be effective, communication deals with creating the mood and turning the audience on to take interest in the messages. It looks at the best timing and the best ways to deliver the messages (most policy-makers have little time to read, face to face interaction might sometimes work better). It packages content in a way that it resonates with the audience. The result of effective communication

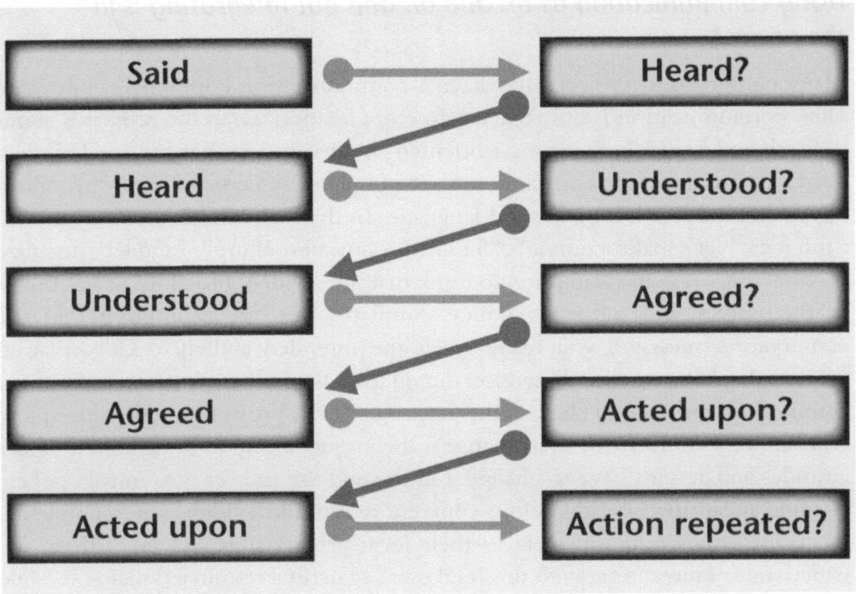

FIGURE 8.1 The assumptions we make when communicating: information does not lead automatically to action[3]

is that they want to know more and ask: "do you have a copy of the full report?" And finally effective communication makes them tell their colleagues and bosses about the evidence based knowledge from your project, research, the recommendations of your policy brief and they may invite you to be on a committee.

In his latest publication, *Do Not be Such a Scientist: talking substance in an age of style* (Olsen, 2009), marine biologist and film maker Randy Olsen argues that scientists should pay more attention to how they communicate their work. They should focus not only on substance or content, but much more on the style of communication:

> . . . communication is not just one element in the struggle to make science relevant. It is the central element. Because if you gather scientific knowledge but are unable to convey it to others in a correct and compelling form, you might as well not even have bothered to gather the information.

To get people – non-experts – to listen to your research or project findings, you have to first stimulate curiosity in them. You do that not through content but through style, humor, spontaneity and personal messages. Only after you have "aroused" your audience are they open for the content.

Using communication as an add on and not integrating it in the project

Many conservation projects may have a communication component, but quite often communication is only taken into consideration when the project is about to finish and suddenly the science oriented conservationists have the feeling they need a communicator to get their message out. They often want to stay in control of the communication means and language. In the best case they come up with a fun idea: "let's make a movie", "let's make a nice brochure", "and let's organize a workshop". It is an assumption to think that just a video, brochure or workshop in themselves will influence policy. Similarly it is an assumption that the conservation messages will really reach the intended audience. Conservation projects that have a policy dimension should from the start think strategically about communication. In each phase of the project cycle the project team should explore who should be informed; what change is the project trying to bring about; whose attitudes and actions have to change if in the end the project is to impact policy. It could mean that already before concrete results and evidence are completed, the team brings policy-makers to their local project sites and gives them the experience of interacting with the "end user" of natural resource policies. It could be quite effective if they analyze options and opportunities together with the researchers of the project team. That approach at least addresses "not invented here" – a common resistance reaction of government officials.

The advantage of integrating communication right from the start is that it saves time and money. When we combine research about the policy context and the key audiences with communication context and the research of current and desired knowledge, attitudes and practices of target groups, it can bring to light which project interventions might be most useful. Thinking of communication when working in the field on a conservation project also saves time and money: often when a project ends and the results have to be communicated, there are hardly any visuals, testimonials or real life stories. Often the people in charge of the project presentation, workshop brochure or policy brief have to go back to make photos and videos or get quotes and stories. Or have to do without and be less effective.

Integrating communication in the project right from the inception phase also means setting aside time and budget for communication. It means planning roughly communication activities and approaches and identifying which external communication expertise might be needed: facilitation of training sessions, events, exhibits and many more. It also means planning of sufficient time to test draft messages or tools with the intended audience, or to involve participants from the intended audience in the preparation and organization of an event.

Not being aware of the principles of systemic change

Influencing policy means dealing with change. It is our observation that psychology and not logic is key to bring about change. Critical dynamics of change are to understand the nature of resistance to changing behavior and the creation of support for meaningful involvement.

In our experience policy change happens when some basic conditions are fulfilled. First of all there should be sufficient dissatisfaction among stakeholders and key decision-makers with the current situation and a real sense of urgency to change the current policies and practices. Second, a powerful coalition should be created among leaders inside and outside the ministry or other governmental institution. Third, there should be a powerful vision for the new policy and clear and agreed goals on what this policy should achieve. Then, stakeholders and policy-makers need to believe that change is possible and a new policy will work. Lastly, the first steps towards the desired impact and short-term wins should be clear.

Conferences, workshops or other communication meant to influence policy are often mainly focused on content or conservation substance, without a clear link to one or more of the conditions for policy change. Without messages that make the explicit link, for example, to creating a sense of urgency, the audience may leave the conference or workshop with the idea that there might be some kind of a problem, but it does not really regard them immediately: "sad, but let's move on". In that case conservationists fail to influence policy.

Without sufficient dissatisfaction about current policies, a policy brief will never do its job. Communicating dissatisfaction has to precede any policy brief. Conferences, life or video testimonials of stakeholders, multi-stakeholder dialogues, a photo report by a journalist on the issue and use of the new media can be effective means to create a shared need for change. But they have to be tailored to each intended audience to realize such attitude change.

Without a powerful coalition to guide change from policy formulation until its implementation and evaluation, not much can be achieved. Communication is the means to form such a coalition, mobilize commitment and "live the change". Networking with influentials to get the backing of key decision-makers in the ministry or (governmental) organization is key to get such a powerful coalition off the ground. Informal lunches and dinners may help. A process of stakeholder dialogues may help to get support from the private sector and Civil Society .

Communication may further help to create a compelling vision for the policy, help to clarify and agree goals and reach of the new policy. Workshops, video-testimonials by protagonists and speeches by the minister or CEO will help to communicate the vision. Expert workshops, conferences and dialogues are tools to shape a common vision, clarify and agree on goals and reach of a new policy.

To mobilize commitment and show change is possible among stakeholders within and outside the ministry or organization, visits to pilot projects or demonstration sites are a possibility. Or testimonials on video or short films can bring 'reality' into meeting rooms.

First steps to implement the policy can be shown in pilot projects and workshops where civil servants learn and practice the changes in their normal routines that are needed to make the new policy work. Investment in communication is decisive whether civil servants will change or whether they will continue to conduct business as usual. Such interaction with civil servants who have to implement the new policy not only helps to make the policy change last but also offers an opportunity for monitoring and learning where improvements can be made.

Forgetting that influencing policy is influencing people

Influencing policy means influencing people. Policy-makers and decision-makers are people. To make them change policies and practices we have to take into account that they are people of flesh and blood. People do not change because they are told to change but because they want to. So the process of influencing people is to get them interested, instil a sense of urgency and a desire to change policy.

It starts with the messenger: who is doing the talking? How likeable is that person? To get attention of the audience and get them really listening, the "how" is more important than the "what". In his book *Silent Messages* (1981), Albert Mehrabian explains – almost at the same time as Rachel Carsson wrote *Silent Spring* – that word choice, content and knowledge only make up 7 per cent of an audience's perception of a speaker. Qualities of how one speaks – voice tone, pitch and inflection – count for 37 per cent of an audience's perception. Visual presence and body language count for 55 per cent.

So when conservationists are talking to policy or decision-makers they should bear in mind that their audience should not be distracted by the way they look. Our advice about body language is:

> When you visit your audience dress simple, wear soft colors, nothing too bright or too dark. Keep your whole attire simple. Behave like your mother taught you. Sit still, do not move too much. Do not cross your arms. It looks arrogant or defensive. Do not wander with your eyes around the room; it may look like you are uncomfortable or defensive. Smile with a genuine smile like you would smile at your grandmother or kids. And before you come to the point, first break the ice.

The same applies when a conservationist wants to get his messages across when talking to a conference of policy-makers or being on camera for national TV.

Influencing people also implies being aware of the psychology of asking something of the other person. And again people respond more easily on the basis of how much they like somebody, than on the basis of the logical arguments and a rational weighing of costs and benefits. This means that in first conversations do not ask immediately for a complete policy change. First establish a relationship of mutual trust and understanding. In this stage it is more about emotion than being rational.

As said before it is important to understand the dynamics of resistance to change. Changing policy and practices means a lot of pain for the ministry or governmental organization. There are always a range of irrational arguments for decision-makers not to think about policy change. Therefore one needs to know in advance what knowledge the audience has of the issues. This helps to avoid irritation when repeating what they already know. One also needs to know what the attitude towards the issue is. This helps to find in advance ways of how to overcome prejudices, fears and resistance. Finally, one has to know the habits, routines, practices and "business as usual" behavior of the audience and the institution they work for. Changing habits and routines are most difficult. And we know from health issues such as smoking that change is mostly not based on information and rational weighing of costs and benefits.

In general it means that for communication to be effective, target group research of each audience is needed to establish what they know and should know, their current and desirable attitude and their current and desirable practices. Communication objectives have to be formulated in terms of knowledge, attitudes and behavior or practices. Often communication objectives are formulated in a rather vague way, such as "to raise awareness among policy-makers" or "to train technocrats in climate change adaptation". Such unclear objectives and lack of "zero-measurement" make it difficult to evaluate the communication interventions. If communication objectives are formulated in the "SMART" (specific, measurable, achievable, realistic, time bound) way and address knowledge, attitudes and behaviour – it makes monitoring and evaluation of the communication interventions easier. Proper target group research in advance is the basis for monitoring and evaluation.

Using messages that do not stick

Often when preparing for a presentation to policy-makers, conservationists focus their messages on the precise content and results of their findings. It leads in most cases to presentations full of jargon, long texts, many details and very little effect. Too often after their passionate presentation about conservation policy high officials tell conservationists: "Very interesting indeed. We are here to help you. Please come to us at any time you need us." This means that the messages so far did not contain an action component. Effective communication is not about what you want to say to people, but what you want them to feel and remember about you when you have left the room. And most of all what the audience is going to do about it.

Therefore initial messages are not about the substance of the conservation issue. They answer questions like why bother, who cares and why? They sketch the bigger picture by providing context, a sense of urgency and first steps to a solution.

Influencing policy and practices means changing the mindsets of policy-makers or decision-makers. An effective message has to resonate with what is important to the audience. You have to present your arguments in their vocabulary. And in most cases that means not in the language of conservation science.

Target group research of your audience can reveal what they already know about the issue. Talking below or above their level of knowledge makes you immediately lose their attention. Not knowing the procedures, workings and titles of their department does the same. The more you know about how the audience views your issue, their interest and understanding, the better you can talk on their level. And again be aware that whatever policy changes your issues imply, they will always be viewed as a hassle as resistance is a normal psychological reflex to any change.

If you know the audience is not enthusiastic, biased or against what you propose, it is best right from the start to name such prejudices and speak to these concerns directly by explaining your arguments. Again messages have to be phrased in what you want your audience to remember. After a meeting with a policy-maker you do not want to be walking away with your audience saying "that was quite interesting" and then have them return to their normal business. In each encounter there should be an explicit message what you want your audience to do: you want their moral support for your project, the use of their networks to test things out, you want financial support, you want their technical staff to attend training, you want a seat on a committee.

What is true for face-to-face meetings is also true for other forms of communication, hand-outs, videos, and many more. They also need to be appealing to feelings and emotions and need to have a clear message of what you want the audience to do. It is always useful to ensure a "plain language" content and have stories with human interest that illustrate your points. Often in a meeting it is useful to have some large photos show casing the issue and tell some stories and personal experiences you had on the ground, how this connects with the priorities of your audience and what they can do about it. This invokes sympathy and interest. That is how you will be remembered.

Avoiding jargon is one of the most challenging tasks for conservationists, as one conservationist testifies: "Every time I wrote about rain in my policy brief, my boss would come and change it with his red pencil into precipitation." Many conservationists will have had similar experiences. But even avoiding jargon is not enough. We have to realize that words have not only their literal or dictionary meaning; they also have strong associative connotations. A word invokes feelings, images, memories and values. People in the disciplines of journalism or advertising know how much words matter. Choose the wrong words in your message, headline or tagline and no one will listen anymore, read your article or contribute to your case. As an example, many policy-makers often do not like the word "green". Avoid the word "green economy"; use "the economy of the twenty-first century". Again this is part of a good target group research.

Applying wrong communication approaches or wrong expertise

Often conservationists are left with the feeling "we produced this nice video or brochure, we organized this conference or workshop and nothing has changed". Analyzing such examples often leads to the conclusion that the wrong communication means were chosen, the wrong combination of means or the wrong expertise was hired.

Communication takes place in many ways: one-way communication approaches (such as mail, print articles, TV, video, PowerPoint presentations), two-way communication approaches (such as visits, lunches, expert meetings, committee meetings, workshops, conferences, demonstrations sites) and participatory approaches (such as joint action research, pilot projects, multi-stakeholder dialogues, all level capacity development).

In general one can say that one-way communication works best when the complexity of the issue is low, the message simple and there is a high certainty that the action leads to the desired outcome. For example, "the text in brackets in paragraph 'x' should be changed from 'y' to 'z' and the brackets should be deleted". At this point, we are in the final stages of the process of influencing policy: the formulation of the text of a new law or regulation. Or the message is: "Don't walk on the grass." This can be communicated through a simple signboard. Other examples of simple messages and clear desired results from actions for which one way communication is effective are posters to stimulate people to come to an exhibition or meeting.

Two-way communication works best when the issue is complex, the message is difficult and there is no immediate guarantee that the (next) action leads to the desired results. Issues such as chainsaw milling are complex, just asking local people not to abstain from chain saw milling will not lead to the desired outcome. The desired outcome can be that politicians become aware that "our country needs a ban on chain-saw milling". That does not mean that a ban will stop the practice. But a first step can be agenda setting and influencing policy. At that stage we are not sure that policy or decision-makers will be convinced of the urgency, goals, and feasibility of the change. Two way and face-to-face communication offers the opportunity for them to be drawn into the subject. They are not told but they become part of the discussion.

Participatory approaches work best when the issue is highly complex, the messages not clear yet and there is yet no certainty that the action will lead to the desired outcome. For example, "Elsewhere conservation is contributing to poverty alleviation and therefore conservation should be also integrated in our poverty reduction strategies." This is where many conservation projects are in the beginning. If influencing policy is not on the "radar" of the project staff right from the start, these projects often stay isolated projects with only some temporary local impact. The knowledge such projects generate is often not tailored to

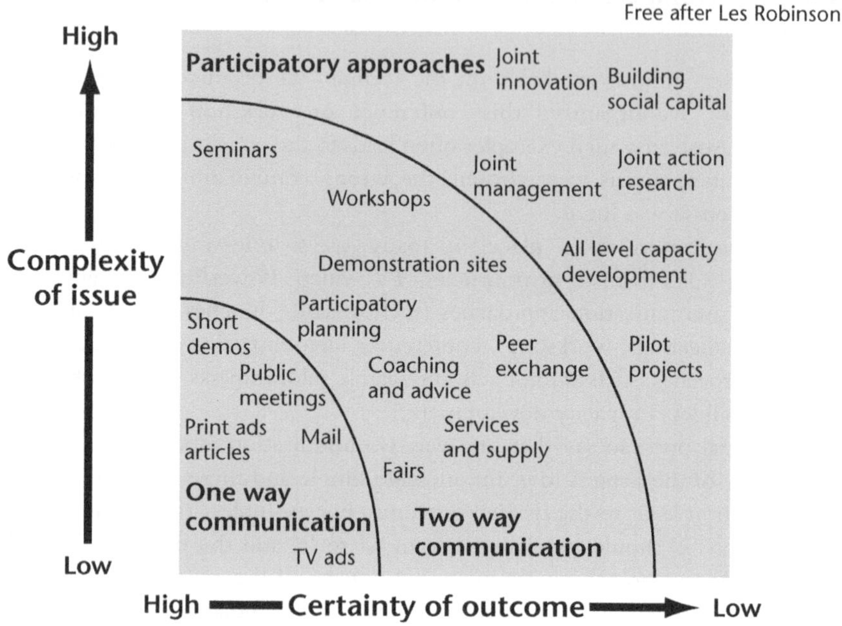

FIGURE 8.2 Objectives and messages are decisive for the choice of the right communication means[4]

influence policies or to up-scale the results. This is another reason to integrate communication right from the start of such action research projects.

Strategic communication means exploring in advance which communication means will help best to get the messages across at what stage in the process of influencing policy. This is part of the audience research. Strategic communication lays out the stages of influencing people from thinking about change to realizing the change. So depending on the phase of influencing a policy, communication focuses on creating different moments of face-to-face communication, often supported by one-way communication means (such as handouts, video-testimonials). It is the right mix and correct timing of communication interventions that leads to success.

For the majority of people the most trusted sources of information are their family, peers, colleagues and friends. Bringing policy-makers into contact with their peers who are already convinced of the change, or have already realized it, is a powerful communication intervention to influence them. Similarly when a policy-maker reads about the issue in the magazine he is used to read, the message often gets across more effectively than through a brochure made by conservationists.

Conservationists often want to stay in control of the whole process of generating, enhancing and distributing the evidence-based knowledge from their projects. In the best case they may hire a journalist to help make a brochure. This makes conservationists extremely vulnerable. A conference where ten speakers give PowerPoint presentations with almost no time for questions is not a way to create the dialogue needed to get policy-makers thinking about change let alone take a first step towards change.

Therefore strategic communication also implies the hiring of the right expertise: facilitation expertise for workshops, conferences, seminars and stakeholder dialogues; PR expertise for events, high-level lunches, visits and brochures; social marketing expertise for audience research, focus groups and campaigns; and media expertise for interviews and getting exposure in the press, on radio and TV.

Forgetting to develop a strategy

Mostly conservationists write reports, maybe have a journalist write up a glossy summary, and in the best case write a policy brief. How to make the relevance of their knowledge known and how to distribute the information is often not thought of. A workshop is organized, but then they often complain afterwards that they did not have the right audience in the room. To bring science and evidence based knowledge to life, a strategy is needed. The strategy is based on what works best for the audience in each specific case.

There are some general strategic approaches including "create word of mouth" as one of them. This strategy is focused on creating the conditions that the evidence-based knowledge from a conservation project becomes the "talk of the town". If everyone is aware of the issue and the need for change, the basic conditions are created for policy change.

"Seeing is believing" is another general approach. It is based on letting the decision or policy-makers experience the reality of the conservation issue on the ground. A personal experience of seeing, smelling, hearing and understanding can often work well, if the event is organized in the right way.

Bringing reality into the room is a third strategy. This can be done by visuals of the actual location of the conservation issue, or showing video-testimonials of stakeholders about problems, solutions and the desired changes in policy.

Another strategic approach is "be good and let others tell it". This approach is based on using certain people as senders of the message who have a high credibility with the audience. This can be ex-ministers, renowned writers, TV celebrities, and others. Such ambassadors sometimes can open doors for the conservation experts that otherwise would stay closed.

Humor is a fifth strategy. Some complex issues are difficult to raise awareness about among a non-expert audience. Invasive species is such an issue. In his latest one-minute TV advertisement, marine biologist and film-maker Randy Olsen uses humor to make the point of the invasive lion fish eating all other fish species in the Caribbean: "Stop the invasion. Eat lion fish! Visit deathtolionfish.org."

Bench marking is often an important strategy to get the attention of policy and decision-makers. They like to know where their country is compared with other countries in the region. This strategy works best when the policy-makers have an attitude of not wanting to be the last one in their region.

Other strategies include the analogy, the inversion of reality. The important thing is to avoid improvising as much as possible. Being strategic means analyzing the issue in advance and understanding the psychology of resistance against change. It also means defining the audience for each intervention and understanding them, and crafting messages that are less about substance and more about what the audience remembers about afterwards, why they should bother, and what they should do. Furthermore, being strategic requires choosing a mix of communication approaches that complement each other. Mostly one develops a communication strategy by thinking through these steps a few times, and each time tailoring the interventions more effectively to the audience.

Sticking to old fashioned prejudices regarding spin, style and PR

Most conservationists consider themselves first of all as scientists and not lobbyists. Even when they want their evidence-based knowledge or science to influence policy, they are allergic to anything that looks like spin or PR. They do not think it is important to have a facilitator lead their workshops. They do not want to associate humour with their findings. They want to be in control of crafting messages. They are the ones that change the word "rain" in a policy brief into "precipitation". They are the ones that argue that the facts should speak for themselves.

These prejudices may be quite valid in the world of conservation science or in the academic world. They are not valid in the reality of decision and policy-making. Scientists may be convinced that their findings are true, that their way of describing the reality is the only logical way of explaining the problems, the solutions and their policy implications. But outside their community "perception is the only reality" and this perception might be quite different from what scientists have proved to be the reality.

What conservation scientists tend to forget however is that in the "real" world of politicians, media and the public, the environmentalists are perceived in a different way. Once they want to influence policy they are perceived in the same way as any other interest group, either from Civil Society or the private sector. And the way they lobby is compared to that of others. Professional lobbying implies using strategic communication, and use of PR, style and spin. This can be different from commercial lobbying, but the principles are the same.

In order to be successful in influencing policy, one needs to refer to a certain degree of spin, style and PR. If not the result is bad communication or no communication at all and bad or no communication also communicates. It communicates that "whatever we say about the issue is in reality not that important for us, otherwise we would have made a better lobby effort like other interest groups (such as from the private sector) do with regard to the same issue". Complacency in overcoming these prejudices against communication, spin and style is a recipe for failure to influence policy.

This does not mean that conservation scientists have to become first class communicators or lobbyists themselves. Once they recognize this they can employ additional expertise. What they have to do though is to overcome their prejudices. They should realize that the usual way they influence policy is not working. They should understand the importance of strategic communication and the steps involved. And in their projects they should make a place for a communication specialist (not a journalist) close to the management of the conservation initiative.

African examples of effective communication to influence policy

Over the recent decades many conservation projects have been carried out in Africa. Some have been successful to influence policy and practices others less so or not at all. We have come across many examples of successful communication interventions towards a range of audiences that are involved in decision and policy-making. Most of the examples are good illustrations of the communication theory, explained in the 'eight errors' outlined in this chapter. As effective communication has to be tailored to each specific audience, examples will be given of effective communication approaches towards:

1. Ministers and high government officials.
2. Technocrats and technical and policy staffs in ministries and para-statal organizations.
3. Members of parliament or elected bodies.
4. Religious and traditional leaders.
5. District level decision-makers and extension officers.
6. Local communities and natural resource managers as decision-makers.

Ministers and high government officials

With the minister or high officials supporting your ideas in general, it becomes much easier to get a place at the table and influence policies and practices. Some strategic communication interventions are: the elevator talk, the opening of an event, a high level lunch, extra attention to the secretaries or personal assistants, a visit, an updating session and feedback. The communication principles of these interventions may also apply to other audiences, such as the importance of investing in establishing and maintain relationships and networks.

The elevator talk

> I am Sindila of the Sustainable Development Think Tank here in town. We are passionate about contributing to the solution of the waste problems. Results from our recent projects may very much enhance the waste management bill you are working on. I know members of your party are very keen to make this bill a success. Would you personally be interested in making an appointment to hear some of the highlights of our findings and concrete recommendations? Who can I phone?

An elevator talk is a one-minute (or roughly 75 words) narrative rehearsed in advance that one uses to make contact and get an appointment with a minister or high official, when meeting such people by chance in an elevator, in the corridors during the coffee break of a conference or elsewhere. It is not about substance, but to introduce yourself (smile and keep eye-contact) and raise interest in your project and finally get an appointment to come and tell more about it. It is important that the audience leaves the conversation with the feeling: "that was a nice person, he has an interesting project that might help the work here in the ministry, I have to remember to tell my secretary to make an appointment as soon as she phones."

The opening of an event

From a letter of a conservation NGO to the Minister:

> Our country has been selected as the only country in Africa to participate in a project on Capacity Building on Biotrade based on the excellent ground work already undertaken here. This issue can also open up new opportunities for a Green Economy approach in line with our Development Strategy, Vision 2030. There will be an inauguration event where international and national sector representatives, as well as several donors, will be present. We would be delighted if you, as representative of the Ministry of Trade and Industry could open the event. We would be delighted to provide you with relevant briefing materials prior to the event.

The opening of an event can be a useful opportunity to inform the high-level decision-maker of an important issue and raise interest in a topic. Giving the Minister the responsibility to deliver the opening remarks does foster involvement. The Minister is likely to want to know what the event is all about and will scrutinize the prepared speech before its delivery. The Minister will put in some effort that the delivered speech is linking in with the priorities of the Ministry and might be alerted to the urgency of the subject matter through this involvement.

High level lunch

The speaker – a former minister of a neighbouring country – ends his short talk on the substance of National Biodiversity Strategic Action Plans (NBSAPs) with:

> As we all can see something has to be done urgently. And it can be done, as I have indicated in the examples from my country. For further reading, the organizers of this lunch prepared some materials on how your country can develop its own NBSAP. They also are organizing a few training refresher workshops for staff of the Ministry. We would be pleased if in your closing remarks you could give a personal reaction to this initiative and indicate to what extent your Ministry will take up on this offer and who they can contact for follow-up.

An informal lunch is meant to raise awareness and get the Minister to a 'breaking position' on more important and intensive follow-up events. The technical content of the event should be limited to just a brief speech – not a lunch full of formal presentations! The lunch should be a pleasurable experience, with good food and a high quality feeling. "If these people can organize a quality event they are most probably also doing a fine technical job" should be the association. The communication strategy behind a lunch with (inter)national "peer" speakers is bench marking. It plays into the psychology of not wanting to be "lagging behind".

Extra attention for the personal secretaries and personal assistants (PAs)

While waiting in the reception or waiting lounge for her appointment with the minister, Nyawira tells the PA:

> May I leave a copy of the report I will present to the Minister with you? Maybe you are interested to read it or you could file it in case the Minister misplaces his copy. I also brought you a special jam made from indigenous fruits from our project, so you can remember me the next time I have to call your office, you know: ah that's Nyawira of the Conservation NGO!

The first entry point to the high level decision-maker is his secretarial staff or PA and through the reception. It is important to build a rapport with these people. If you are likeable and understanding, it is a lot more likely that they will go out of their way to accommodate you next time you are seeking for an appointment.

Being likeable and thoughtful – for example, bringing a small thank-you from the project – is a thoughtful gesture. Of course it is important to ensure that such a gesture is not misconstrued as bribery. It should be emphasized that such gifts only work with corporate merchandise or products as is the case here: jam from our project. It is important to be kind, jovial and keep a smile even if there's a lengthy waiting time or a long awaited appointment has to be cancelled on short notice. Sometimes, to have a successful meeting, one has to persevere positively with the effort of getting the appointment. However, it is of course perfectly fine to request to be notified if there is a change or cancellation. If you do that in a friendly manner, it is more likely that you will get a call in good time next time around.

A visit to meet on the substance

When seated in front of the Minister's desk in the newly created Ministry for Drylands, Abdi is told he has a twenty minute appointment as afterwards the minister has other pressing engagements. Abdi opens:

> Minister, thank you for your time. We are passionate about dryland conservation. And you know, drylands are often dismissed as investment deserts. However, our studies show that the reality is very different. Drylands produce roughly the average per capita GDP despite receiving far below the average per capita public expenditure. In fact drylands produce a higher return on public investment than more humid areas. Policies that do not recognize this and invest in new ways of land use, run the risk that such alternative land uses that may seem attractive, in the end represent a cost that far exceeds the benefits. We are organizing a workshop to define a study to research this in more detail in the North of our country. We would very much like your moral support for this study and to have experts from your Ministry participate in the workshop. This way we can ensure the study will address the right questions and will produce results that can add value to the policies you are developing.

The speaker attracts the attention by using short sentences, the analogy of 'investment desert', touching on the core business of the new Ministry and coming immediately to the point.

After the conversation the Minister can clearly remember the message: the promise to generate knowledge how to avoid future costs of wrong land use policies. And the request for concrete action by him: his blessing for the project and his experts to help define the study. As the study is financed by external sources, the study might prove to be an unexpected opportunity for a Ministry that is structurally under-financed. This makes it even easier to give a positive reaction to the request.

Updating session for high level policy-makers

This is an extract from the invitation to the permanent secretaries, deputy secretaries, and the directors of departments of several ministries:

> The National Forest Authority is organizing a special meeting on REDD in partnership with IUCN. The Regional Director of IUCN and Forest experts from IUCN headquarters in Switzerland will come to our capital especially for this meeting. They will explain to us the opportunities offered by this mechanism from the international climate change negotiations. They will particularly address how our country could benefit from the special funds that will become available. You will have an opportunity to explore how your ministry or department can prepare itself to become eligible for these funds, so you can instruct your staff accordingly. The meeting will take 2 hours at the Hotel Imperial and the hotel will provide lunch and tea.

When high level decision-makers agree that a certain course of action is important – in this case "readiness for REDD" – the relevant technical advisors and mid level decision-makers will be stimulated to update themselves on the details and the various steps that have to be undertaken. The need for this high level meeting was articulated earlier by a few technical staff that were present at the first training meeting on REDD. As one expert observed: "In policy influencing, communication never ends. Once you have a willing ear at one level, you need to see how to get to the next level to make real impact."

Positive feedback

Solange, who was responsible for the invitation to the high level lunch on behalf of an influential local NGO, is drafting a letter to the Minister after the event:

> We highly appreciated your presence at the high level lunch last week. Learning from the experience elsewhere in Africa as presented by the Honorable Minister from Mozambique, we would like to inform you that our NGO can support your ministry in setting out a planning process for developing our NBSAP. Following up on your closing remarks at the event, requesting us to assist your ministry in such efforts, we have prepared an initial concept and we would like to discuss this with you. We also have prepared the first training events for your technical staff and aim to commence with these sessions within the next month, preferably after discussing the proposed NBSAP process with you. We intend to contact your office shortly to discuss a date for a meeting with the Minister.

The importance of feedback and follow-up is not to be underestimated. It is not just simply good enough to have the representative at the one event, but engagement into a long-term dialogue and process is desirable. Our communications must be designed taking this into consideration from the onset. A brief thank you letter to the Minister can be used to communicate such follow-up and briefing of next steps. After the high level lunch, which was designed to raise interest, a more detailed information and substance discussion can take place.

Technocrats and technical and policy staffs in ministries and para-statal organizations

To get technocrats interested in your information, it is important to make it clear how this knowledge adds value to the plans and priorities they are working on and or how it can reduce costs or tap into new financial resources.

Participating in committees

Panduleni is an environmental professional and works as a consultant in a local expert office. She shares how she has become a member of the national climate change committee (NCCC) in her country:

> For years I have worked as a consultant to our environment ministry. Recently I presented a critical resource paper on how climate change may affect the national fisheries sector, a key to our national economy. The Permanent Secretary who attended the meeting was impressed by the candid and technical presentation and requested that I would be invited as a member of the NCCC.

Joining important committees, either by invitation or by assertive self-identification, can facilitate powerful opportunities to influence policy. Getting onto a committee can sometimes be influenced by us. Sometimes a colleague at the office is better positioned to receive an invitation, maybe because she is more senior or politically better connected. We can then use this person as our spokesperson. But often we can use our own potential by demonstrating technical soundness and professionalism. We can also inquire proactively if the committee would like to benefit from our inputs, availing ourselves potentially at no cost. It may be that a development partner or donor has some say in who would be on a committee. If we have a good working relationship with such role players we might benefit from their support in getting recommended.

Using your seat at regular committee meetings can be a related opportunity. At the agenda item 'Any other business' of the regular Sectoral Group Meeting of technical advisors of various departments, Barbara takes the floor:

> Thank you, Chair. Many of our discussions we usually end with the conclusion that a lot has to be done in the environment sector. However,

> we immediately add: unfortunately we have no funds available. It is this issue I would like to address. For the forest sector, REDD offers a window of opportunities. REDD stands for Reducing Emissions from Deforestation and Degradation. It is part of the international negotiations to combat climate change. The question is how can we qualify for these funds? My organization is willing to organize a half-day meeting on how the plans of your departments can benefit from REDD. My question is who of you is interested to participate?

Short and non-jargon sentences keep the attention. The message plays into the major concern of the meeting about the costs of environmental plans. And it contains an immediate action perspective: come and learn about REDD. The result in this example was that 17 technical advisors came to this first workshop.

Using a professional updating session to influence a policy process

Professional updating is a term that recognizes that the target audience is already composed of professionals that have undergone specialist training in their profession. This is a relatively new terminology that is being introduced in the field of conservation and sustainable development. The sort of target of professional updating could be:

> At the end of the training the participants indicated a better understanding of REDD and the Readiness process that is on going in the country. Participants committed to play their part in supporting and expediting the REDD readiness process. They tasked the National Focal person to update them on progress. However, the participants indicated that there was need to organize a top policy training/presentation on REDD for high level decision-makers to enhance the process.

Professional updating sessions are not only a means to get the messages on the substance out but they are also an opportunity to explore with the audience the most appropriate next step. To make these sessions function optimally for adult learning and two-way communication, it is advisable not only to hire external facilitation expertise, but also to prepare the session with one or more potential participants from the intended audience. They also may help in identifying and inviting the participants.

Peer reviews and joint report writing

James says:

> We engaged in an interesting and interactive writing and peer review process on a project on financial and investment flows for adaptation in

> the agriculture sector this past year. The study was carried out by Government and designed to include private sector experts in its process. Through the interactive technical work sessions and peer reviews, we were able to help direct the final report and ensure that our technical expertise was included and considered.

Including technocrats from the public sector as co-authors – or setting up joint writing teams can be a powerful way of establishing joint working teams on critical technical issues. We can design our projects to ensure that relevant stakeholder become co-authors of important policy informing research work and support studies. Ownership as well as technically sound understanding can be created in such a manner. The communication strategy behind this intervention is based on the principle: "Teach and they forget; demonstrate and they will remember; engage and they will do". This intervention works well when relationships of mutual trust and understanding are already somehow established and the audience already is aware of the answers to the question "why bother?"

Publications and meetings within their field or sector

The SADC Regional Water Programme wanted to reach technocrats and policy-makers from other sectors to adopt the concept of integrated water resources management (IWRM). Common practice had been to send them publications on IWRM and invite them to IWRM workshops. There was low participation in the workshops and it was not clear that the publications were even looked at.

The SADC Program changed its strategy and published articles on IWRM in the newsletters and magazine published by the respective sectors. Technocrats always read the publications from these sectors and putting the articles in their newsletters and magazines maximized the chances of being given attention than sending them newsletters and magazines that are alien to them. For example, in order to reach the technocrats in the energy sector, the program published an article: "Water and Energy Security in Southern Africa" in *ESI Africa* – a magazine for the electricity supply industry.

The Program also negotiated a slot on the agenda for a presentation during a workshop by different sectors. Obviously, technocrats prioritize workshops within their sector and taking the presentation to them was more effective than inviting them to a meeting discussing water. For example, to reach the energy sector, the program made a presentation on "Hydropower and integrated water resources management" during the Hydropower Africa Conference, which brought together utility managers, engineers, generation managers and renewable energy managers from the energy sector.

Members of parliament or elected bodies

Members of Parliament or elected bodies can be important policy agents and it is useful to interest them with your concerns. Either they can become visible champions for your course, or they must be convinced of the conservation interest to sway their potentially conflicting interests. It is important to identify areas of interest and hot issues, to which a larger policy topic can be linked. Often this audience may not be fully aware of possibly interesting topics that they could use for their own work.

Raising interest – the first introduction to the issue at hand even at a public meeting

Thandi met the representative of the leading party in her constituency at a pre-rally event. She had the opportunity to raise a critical issue in the plenary and to follow-up with the candidate shortly after the event. In the plenary, Thandi raises the issue:

> We have been able to review the commissioned EIA report for the new cement factory that is planned at the outskirts of our regional town. Although we understand that numerous needed jobs will be created, the associated health risks and environmental concerns may outweigh the potential benefits of the investment. What is your reaction?

After the event Thandi had the opportunity to speak to the candidate personally:

> The issue on the cement factory really is of concern to me and our NGO. We are involved on a number of issues that relate to the follow-up on EIAs and compliance in our country, and we would like to see our EIA legislation to be improved. With the upcoming elections this might be an important issue for you to get some background information. When could we come by your office and give you a detailed briefing?

Similar to the elevator talk, it is important to raise the interest of your target person to your concern or project. People listen to what appeals to them. In the context of elected representatives your message needs to be associated with what is important to your audience – and their constituency. Choosing a small topic that raises direct stakeholder concerns is a clever way of raising interest – the full project content or policy concern can be introduced at a later stage. It is important to also seek out opportunities to meet the policy-maker, in this case the parliamentarian, councilor or other elected representative. There are usually strategic events, at which such people can meet. Although they do not usually want to necessarily be seen as political lobbyists, taking a slightly politicized approach to influencing policy is absolutely legitimate – and often successful. It is of course critical that we do not position our topic in a manner that it becomes strictly party political and might be sidelined if we were "betting on the wrong horse".

"Updates" – short policy issues papers

Ishmael, a representative of the environment ministry recalls:

> An idea that has been extremely useful in our country has been the production of so-called "Updates", short and focused policy issues papers on current environment and development issues in our country. These Updates are distributed monthly in Parliament. They raise the awareness of this audience about critical topics that may be important to their constituencies – and the next elections. We have been working together with one of our local NGOs in preparing these policy papers, and are excited about this partnership. We designed an impact monitoring element and have received wonderful reviews for our product!

The importance of this communication intervention is that it is a way of networking that helps to create partnerships for change in the organization and beyond. Providing technical information to parliamentarians and other elected bodies can be very important. Raising awareness about technical issues that could form part of a political debate can also be necessary to start influencing policy. It is important to design the written information in a manner that the information is succinct, relevant to the target group, raises interest and, very importantly, that its presentation is attractive – and professional.

Organizing an advocacy event

Setting up an advocacy event either as NGO or in partnership with a relevant government institution can be interesting for this type of audience. It is important to either organize such an event in the margins of an already ongoing event at which the parliamentarians will be present or sent out specific invitations. Jocelyn, representative of a national consumer association provides an example:

> In our country the Environment Ministry always plans a specific activity for World Environment Day. They invite any environmental organization or interest group to have stands or activities associated with their programme. We usually take advantage of this offer, especially as the Minister personally sends out invitation to parliament to be present at such an event. This year we had a stand on the danger and potential of GMOs

> in our agricultural sector, as our country is currently developing a national Biosafety policy.

We organized a public debate during the day, with selected key speakers, including three parliamentarians and regional councilors, and had a very attractive stand with plenty of outreach and information materials. We prepared a special policy issue paper that was distributed to parliamentarians coinciding with World Environment Day. Several important representatives of our key target group attended the event and later on asked for specific in-depth briefings on the subject matter.

It is important to understand that communication expertise is not only needed to make an attractive policy issue paper, but to organize the event in a way, that it creates the right mood for participants to be open and willing to discuss new issues. It starts with the message of the invitation: how inviting is the tone of the letter? Does it answer concretely the questions "why bother", "what is in it for me" and "what can I expect"? Communication also has to influence the program: "is there enough space for informal contacts and relationship building", "is there room to explore potential doubts, prejudices, resistance", "is there a possibility to ask questions and learn from your peers"?

Religious and traditional leaders

When approaching religious and traditional leaders it is very important to follow the local culture and customs, and to observe protocols to ensure that a positive communication process can be engaged in. Usually it is useful to have a local person or a member from the religious denomination on the team to ensure that relevant background information is assembled before the communication strategy is being developed.

Asking for an audience and engaging traditional leaders

"It seems to be the most trivial thing to do, but still we often do forget or do not take out the time to actually go and meet with the traditional leaders", says Talamondjila:

> In our project on establishing communal areas land and natural resources rights, right from the beginning we engaged with the traditional leadership in the pilot area, to discuss whether the policy directive from the central Government was in line with local visions for the management of commonage areas. Through my father, who is a local headman himself, we got an audience with the king of the Oukwanyama tribal authority.

The entire project team, national and international staff, came to the event and we brought along a beautifully hand carved walking stick as a customary present – and to make the connection to our project, which is promoting the valorization of natural resources in communal land areas. We were at the palace for the entire day, following customs and bringing up technical discussions around our topic of interest. It became apparent that the traditional leadership is not in line with aspired policy changes by the central government. We identified many critical issues, which we are now following up on by linking the traditional leadership more strongly to the ongoing policy process.

What Tamondjila recounts highlights several important communication issues when dealing with traditional leadership. First, it is important to acknowledge the central role such leaders have in policy processes. Second, it is important to find the right entry point to get to know the communication partner. Often it is useful to have a team member on board who knows the local customs and who can establish the contact with the leadership. An audience must often be requested well in advance and the request must follow specific customs and procedures. Engaging the traditional leadership in a discussion about the technical issues at hand and of relevance to the local community will not only lead to the development of more informed policy processes, but additionally pave the way for local level implementation of provisions that are well aligned with local needs.

Taking a walk in the wild side – creating empathy and establishing a dialogue

At the last village visit, Pierre, who is working in a project protecting the Mountain Gorillas, invited the local traditional leader to a site visit:

> Let us go out to the forest and you show me directly what your concerns about conflicting land uses between your village and the conservation area are. I will also take the opportunity to show you some of our conservation results and benefits that have been created for community members.

Seeing is believing. Inviting a local leader who might be apprehensive towards a specific conservation effort to see the benefits in situ can be very powerful. Also inviting for a dialogue, hearing out the concerns and understanding the resistance to an intervention can help to engage the policy-maker.

Walking together is a powerful communication intervention. The psychological effect of not facing each other but looking in the same direction makes a dialogue and mutual understanding much easier. You do not even have to talk all the time and when talking, you can also discuss what you see, what you feel, your previous experiences, and much more. Walking breaks down physical barriers too: you may even touch each other, while negotiating a difficult passage. Walking brings people closer together. In general when official communication on the substance comes to a halt, the way forward is to go for a walk.

Making the "champion"

Mohammed shares his experience from a recent idea for reaching out to traditional leaders to get their engagement for land care approaches:

> We established an idea to identify "land care champions" amongst our traditional leaders. Although not just limited to this target audience, we got some good buy-in from this target group. We now have some 50 traditional leaders throughout our country serving as "land care champions", carrying the flag of land management best practice.

Identifying "champions" for a topic is another good way of creating ownership, and establishing a communication platform and engagement. The identification of such a distinguished group of policy leaders must, however, be well prepared and the process must be well executed, to have some credibility and not to create ambiguity as in "why is she a champion and not me?" Nominating and positioning traditional leaders as champions can have strong ripple effects amongst the community and create strong policy impacts.

District-level decision-makers and extension officers

For local decision-makers it important to get (unbiased) evidence of what is really working at the local level in order to achieve the objectives of their plans, new bylaws or to change procedures. In rural areas literacy is very limited and even among literate people there is a very poor reading culture. Face-to-face communication is therefore the best way to influence policies at the local level.

"Demonstration" meeting

During one of her regular visits to the district officer, project officer Sophie asks:

> The IUCN Livelihoods and Landscape Project has worked now for quite some time with several communities in your district. They really have taken an interest in changing their livelihoods and resource management in a more sustainable way. The most active representatives of these communities would like to present their experiences to you and your staff and discuss with you the consequences for the bylaws. When could we best plan such a meeting?

The power of this communication intervention is its personal character. Setting the stage to get first hand information from local people affected by policies is an experience that potentially can have a large impact. Even more powerful is experiencing the reality on the ground by a field visit. Regular communication with local decision-makers is key to influence local decisions on policies and practices. The minimum is to keep district officers informed. Even better is to engage them in the project right from the start, so they feel like the co-owner of the results. Interacting with communities in a constructive way is often new, but very rewarding for many local decision-makers. Facilitating such dialogue requires a neutral position. Often project partners from outside the district can bring such qualities in. Communication has to be focused on building and strengthening of relations, creating an atmosphere of mutual respect and trust and overcoming prejudices and other psychological resistance mechanisms. It must be noted though that these processes take time.

"Demonstration" field visit

> One communication approach that has worked wonders with us is to organize demonstration visits to actual field sites with water affairs extension officers. Although these officers are supposed to facilitate community outreach and serve the local people, they often are bound to their offices due to a lack of vehicles, fuel limitations, or a lack of know-how of how to engage with their client group,

says Fatima, the coordinator of a water basin management project. "By taking the extension officers out to demonstration sites and organizing stakeholder meetings with other service providers and local level water management committees, a practical communication platform is created, allowing for important exchanges of experiences."

A field visit is based on the communication strategy of "seeing is believing". It is therefore important to invest in the preparation of such a visit to make sure that an optimal life experience of the policy issue is created. This often has to be balanced with the informal and fun aspects of an excursion. The latter are important as they open people up for enhanced learning, but should not dominate too much. This is especially the case with excursions overnight. It often helps to engage a member of the target audience in preparing the field visit. This way it is easier to set the stage in a way that it optimally connects to the current and desired levels of knowledge, attitudes and behavior of the audience. This implies – if the budget allows – to engage a communication, facilitation or PR expert to plan the visit in a way that learning addresses all these levels in a complementary way and does not only concentrate on substance.

Training workshop

Excerpts from an invitation letter sent to various district level ministerial offices:

> The National Climate Change Committee has commissioned the organization of a district level workshop on the topic of building a climate change resilient health sector. Climate change projections for this district indicate

> drastic changes in the climate within the next three decades. Increases in the spread of human, animal and even plant diseases and pathogens due to higher temperatures and increased seasonal rainfall will pose significant risks. Your sector will also be affected. We would be delighted to have you as a key participant in a unique training event. Please contact Mr. Mulunga by the 20th of May for confirmation or queries.

The importance of this communication intervention – if it is well structured and facilitated by professionals – is that it focuses on social learning in a group. Group changes in attitudes and behavior are more sustainable than individual changes. The training therefore has to focus not only on substance and knowledge, but also on attitudes and the change of practices, addressing potential resistance and obstacles in the institution and ways to overcome them. Just technical training will not affect behavior and habits, let alone lead to policy change. As with the field visit, part of the success is in hiring the right expertise here: an experienced facilitator.

Giving a project task to the district office

As part of the development of a new conservation programme, the hired consultant visits the district office of the Forest Agency, who is coordinating the preparation mission at head office level. She is sitting in a pre-planning meeting with the district officials:

> It is very important to invite a wide range of district level stakeholders to the intended program planning meeting. It is important to collect many inputs into the planning process, to identify critical policy gaps and ways of how these can be addressed through the envisaged programme. A stakeholder involvement plan could be sketched out at the meeting, including on communication channels and mechanisms during project implementation. Could your office organize and host such a planning workshop? Do you need help with planning the agenda?

Actions speak louder than words. Giving the district authority the responsibility to set up the multi-stakeholder meeting reinforces their leadership role in the policy planning process. Such reinforcement of authority – and responsibility creates ownership and engagement. Encouraging multi-stakeholder involvement from the beginning of a policy planning process and communicating on the policy is important to facilitate participation. If needed, planning support could be offered, opening a valuable entry point for policy interaction. Meaningful engagement is a communication strategy to change attitudes and habits. It takes time and investment in supporting this process, in order to avoid risks that the financial support is accepted, but "business as usual" continues.

Local communities and natural resource managers as decision-makers

There are numerous examples of how governments are attempting to engage in truly consultative policy development, while they usually battle to get the policies implemented to the ultimate stakeholder level. Communication is a major challenge and we may need to employ multiple strategies: intense face-to-face interactions for depth and broad scale campaigns for breadth.

Participatory policy development – communicating lessons learned from local case studies

A much heard frustration is:

> In the past years we have learnt so many lessons from our local level natural resource management approaches. I would just wish we would have a more systematic way of analyzing them and including them into our policy review and policy development processes.

"We would want to have the views of the local people reflected in our policy development process, but it is so difficult to reach out to a representative stakeholder group" is another common complaint.

The strategy behind this approach is that good communication is at least 50 per cent listening and 50 per cent talking. Communication to change behavior is more successful when the audience is properly listened to in advance. The best results are realized when the audience is engaged in a meaningful way in the process as a whole. The more they feel they own the problem, the more they may become the owner of the solution. Communication is focused on creating the mood, interest and abilities of the audience to get into the driver's seat.

Facilitating local level policy inputs through establishing communication pathways

"As a communication officer at our NGO, I have been responsible for designing a communication strategy to generate local level policy inputs for our participatory poverty assessment", says Ndlovu. "The important thing was that communication was not only an afterthought, but before the policy consultations went ahead, sufficient time and resources were dedicated to its preparation and its execution."

Communication planning is a critical ingredient for successful policy processes. Especially in policy processes that seek the inputs from the local level individual, sufficient planning, adequate resources and dedicated staff are needed to put communication at the heart of the project intervention. Establishing dialogues and communication opportunities is an important policy tool. The communication strategy behind this approach is opening up and strengthening relationships with existing and new networks, needed to build up an alliance for change. The communication is focused on showing that change is possible, that ideas on how to do it can be shared, that you are not alone, and that you can learn from each other and build confidence.

Using a communication campaign to spread the message

At a meeting at the Drylands Agency, Verity raises a point about "spreading the message":

> We must carefully strategize what type of communication or outreach is useful in ensuring that the policy directive on using fuel efficient stoves to conserve our woodlands from deforestation will be heard by the relevant people. What are our best outreach strategies and communication mechanisms?

The communication strategy of campaigns is often based on "using multipliers" and "intermediaries". People who spread the message for us through their networks (agricultural or health extension services) and channels (radio campaigns) are often seen to be useful to even reach people in very remote areas, and many people use radio as their key source of information and entertainment. Most importantly any communication strategy should be based on a gender analysis, as women may play a key role as intermediaries and change agents.

Conclusions

Most conservationists think of communication in terms of means and media: press releases, brochures, videos, and many more. Often little or no thought is given to which audience they need to reach, what change in knowledge, attitudes or behaviour communication should aim for. And what the most effective messages should be. Often conservationists focus their communication more on substance and less on style. They forget however that our human brain works differently. Our minds are ruled by two different systems – the rational mind and the emotional mind. Mostly the rational mind only takes over, after the emotional mind has given the green light.

Good communications is about setting the stage, creating the mood, crafting the right language and images, finding the right time and right means to deliver the messages. And, most importantly, clearly state why the audience should bother and what specific actions are expected from them. Only then can one think about the most effective communication means.

Proper attention to effective communications as a support to influencing policy starts with analyzing the role of communication right at the start of the influencing process. Key for success is the preparation and planning of communication: who is most important to communicate with. Invest in researching what the existing and desired knowledge attitudes and practices of the audience are. Pay attention to the question whether the emotional mind still is in control or whether we can already come up with rational arguments. Think of messages in terms of what we want our audience to remember about us and our issues and what we want them to do after we have left. Explore what strategy would be most appropriate to realize the desired change in knowledge, attitudes and behaviour. Only then should we think of communication means and media.

The African stories of effective communication to support a policy influencing process show how different audiences are, and how to tailor communication approaches to these audiences, depending on the stage of the policy process, the cultural, political and social context. They also show that influencing policies is influencing people. Therefore face-to-face communication is often the best approach, especially in the initial stages of influencing. The wide variety of successful approaches further offer a range of opportunities to integrate communications more strategically right from the start of a policy influencing process. They also prove that success is partly based on a proper budget for hiring the right expertise for the policy influencing process. Conservationists are mostly much better in the role of resource persons than in the role of lobbyists or facilitators of a policy change process.

9

STRATEGIZING TO INFLUENCE POLICY

A framework for practitioners

Jonathan Davies

Policy revisited

One of the objectives of this book is to challenge the sometimes simplistic interpretations of the term "policy", in order to demonstrate the great diversity of opportunities for non-governmental (and even government) actors to influence government decision-making. Considering the significant rise in donor funding to encourage policy dialogue, a broader understanding of government decision-making processes will contribute to more effective engagement of non-governmental actors, and should help those actors to exploit their individual strengths more effectively. Chapter 1 discussed the divergent interpretations of "policy" in order to pre-empt ambiguities and contradictions that were inevitable in subsequent chapters, and the point is revisited here.

As the definitions in Chapter 1 (Box 1.2, p. 10) illustrate, the term policy has diverse interpretations of varying precision. At one extreme the term can be interpreted as a clearly identifiable standard or procedure that determines the course of action of a government, while at the other extreme the term may simply mean the wisdom that helps people to make decisions. Consistent with a number of sources (Young, 2005; Court *et al.*, 2006), this concluding chapter maintains a broad interpretation of policy as simply "a course of action", focusing particularly on the action of government at both local and national levels, and including declarations, formal plans and documents, as well as actions, practices and norms.

It is helpful to maintain a broad interpretation of policy in order to allow more flexibility in trying to influence government decision-making. However, it is also important to understand more clearly what one is trying to influence, and strategize accordingly. For example, the approach to influence a specific piece of legislation will be very different from the approach to influence local government planning and budgeting. Decision-making in government can be highly informal, and even arbitrary, and if an agent wishes to influence this they may need to act in equally informal ways. Understanding the often chaotic and political nature of such decision-making is crucial if an agency is to understand where it can realistically expect to have an influence. It is also vital in the formation of alliances, since there may be trade-offs between credibility, legitimacy and overall influencing power. Institutions that pride themselves on producing high-quality science, for example, may be reluctant to engage in the political struggles that can surround decision-making, but may partner with agencies that are more engaged at this level and who can benefit from improved scientific and technical arguments. The formation of strategic alliances is revisited later in this chapter.

Clay and Schaffer (1984, cited in Sutton, 1999) wrote that:

> The whole life of policy is a chaos of purposes and accidents. It is not at all a matter of the rational implementation of decisions through selected strategies . . . Policy-making tends to become the mystique of elites . . . and these elites are separated from people.

The general public are frequently considered to be poorly informed, apathetic and not worth engaging in decision-making processes, and at the same time vested interests in government may often remain opposed to meaningful consultation that could undermine their own position. This position is constantly changing and there are signs of improvement in a number of Eastern African countries, where political environments have become somewhat more open to input from Civil Society in recent years, and a greater range of Civil Society actors are now engaged in policy processes. The legitimacy of citizen input to policy-making is gaining some strength, reflecting the pluralist model of the State in which citizens have as much right to advocate for their interests as businesses and other interest groups (Court *et al.*, 2006; Tyler, 2006).

Despite greater engagement, there are fears that the opportunities for influencing policy are not being effectively exploited. In some cases this may reflect lingering constraints and reluctance from decision-makers to become truly open and accountable, but it also appears to reflect weaknesses in the capacity of Civil Society Organizations and research institutions, failure to collaborate effectively, questions over credibility and legitimacy, and poor use, or even dismissal, of evidence as a tool (Court *et al.*, 2006). Processes of policy consultation are inevitably imperfect and prone to manipulation by those seeking a quick, or perhaps biased, outcome (Tyler, 2006).

Non-governmental actors engage in policy dialogue with government to promote adoption of experiences and lessons in policy formulation. Policy engagement can also be carried out for the sake of empowerment: as something of intrinsic value in a rights-based approach to development. However, the use of evidence should not be overlooked in either case, whether to ensure that policy options are credible, practical and proven, or simply to ensure that decision-makers respect and continue to dialogue with Civil Society. Nevertheless, genuine influence of policy through purely evidence-based approaches can be very limited. A recent study of the African Energy Sector (JBIC Institute, 2006) found that policies were influenced by only a small portion of available research, and the influential research was almost exclusively commissioned by government itself, a risk that is discussed in Chapter 7 in this book.

Governments thus maintain particularly tight control over setting the policy agenda or goals, with "consultations" at this stage often exclusively at cabinet level. Wider consultations often occurred later in the policy process, but the view amongst researchers was that by this stage the policy-makers had already made up their minds as to the outcomes and the consultations were merely cosmetic. Perhaps even more worrying was the fact that, far from being objective and open, policy-oriented research was found to be driven by the demand from policy-makers (JBIC Institute, 2006). Chapter 4 in this book has illustrated the importance and the possibilities for engaging with elected representatives, particularly members of parliament, in order to influence policy at the initial stage of setting the agenda.

By recognizing the unpredictability and sometimes opacity of government decision-making, it is possible to take a more ambitious approach to ensure influence, and to avoid being constrained by notions of how we would like policy processes to be. Decisions that are made by government are likely to be subjected to a range of influences, almost certainly from lobbyists with divergent or polarized goals. Indeed this may be the fundamental reason for policies: if there were no disagreement over which course of action was appropriate, there would be less need for a policy to guide decision-making. In this regard, the development of a policy document or law is only one step at which influence will be brought to bear, and lobbying will continue right through to the final interpretation and implementation of policy, or the lack of it.

Although Chapter 1 challenges the simplistic notion of the policy as a neat, cyclical process, such conceptualizations are helpful in understanding where some of the entry points are for influencing decision-making. In fact, if we take the loosest definition of policy as relating to the attitudes and behaviour of people (particularly within government), it is highly relevant to think of policy-influence as a process. It is a process of informing, educating and challenging misconceptions, and in some cases it may be part of a wider process of social change. As Chapter 1 states, it is important to keep in mind that influencing "policy" is a means to an end rather than an end in its own right: ensuring a particular decision is made in government could be temporary if there is not an ongoing process of influence and informing people in order to sustain the change.

Contradictions in policy

Far from being a neat and systematic process, policy-making can not only be chaotic and opaque, but can often lead to formulation of contradictory policies.

Within a country, different policies may conflict or contradict each other, both between sectors and even within the same sector, as illustrated in Chapter 1. Policy-making does not necessarily demand a single objective or a single instrument, and multiple objectives and instruments are often found, although this can lead to, or may stem from, a conflict of interests (Torjman, 2005).

As a result of conflicts between policies and policy goals, government actors still have to make judgements over which course of action to pursue: over which policy to prioritize. Decision-making is therefore clearly not a simple case of following the rule book of a neat set of policy documents, and decisions may be made based on interpretation, at local or national level, according to goals of different ministries or departments. The "policy" that guides this decision-making may be unwritten, but is evidently worth influencing.

Identifying the policy-makers

Decision-making within government bodies is influenced by many different actors in different ways, depending on the nature of the decision, the degree of access of different actors to people within government, the level of motivation and capacity of different interest groups, and the nature of existing conventions and practices within government. Some actors may be more effective at lower levels of government – for example, community leaders might exert greater influence over district-level budget allocations – while others may be more effective nationally. Some groups of actors may be more influential in the process of setting government's policy goals – for example, parliamentarians in some countries – while others are more influential in determining the methodology of policy implementation. Understanding the relative strengths of different actors in different types of decision-making can help to make advice or advocacy more influential, as all the chapters in this book have highlighted.

One group of influencers can also exert influence over another group, as for example when a local constituency persuades their elected representative to raise a particular question in parliament, or where private business interests fund scientists to gather data in support of a particular policy option. This interaction depends on the degree to which the different interest groups are informed of a decision-making process and may be influenced by the extent of government consultation, although some lobbyists may find a way to influence decision-making regardless of whether or not open consultations are carried out.

Some researchers may consider, or desire, that the link between research and policy is a linear process: where science is conducted independently and findings are generated and then conveyed from the "research sphere" to the "policy sphere", leading to an impact on the decision-making process. In reality, as Chapter 7 demonstrates, the policy process has been shown to be much less predictable than this and researchers are often much less independent than they (or others) would like. As a result, debates now focus increasingly on a more dynamic approach that recognizes the complexity of decision-making processes and the two-way process between research and policy that is influenced by complex relationships and sometimes conflicting knowledge. Researchers frequently question what it is that makes some research more influential than others, why some research is ignored despite its apparent value, and how they can make their own research more influential and noticeable (Start and Hovland, 2004). In Chapter 8 of this book, the authors emphasize this point and stress that quality of communication is probably of greater importance than quality of research in terms of the likelihood of adoption.

One way in which external agents can contribute to policy dialogue between different interest groups is through convening relevant opportunities, which in some cases may be endorsed by a government department or ministry that simply lacks resources to convene such dialogue itself. Encouraging such dialogue between opposing interest groups can be productive and is critical in opening up new space for marginal interest groups, such as indigenous peoples, rural communities or small-scale producers. The greater the polarization of views and opinions, the more important it is to create a consultative process in order to reach consensus and compromise. Rather than avoiding confrontation, if policy is to be instrumental in reconciling competing interests, processes of policy formulation need to actively engage divergent views. Obviously members of particular interest groups may be opposed to this and are likely to prefer to monopolize the decision-making process where possible.

Policy engagement for communities can be empowering and can be a goal in its own right, and not only a means to an end. Policy engagement can contribute to increasing accountability, exercising citizen's rights, promoting an enabling political environment and ensuring the voices of the poor or marginalized are heard in policy discussions. For this reason policy-engagement can be an important element of a rights-based or empowerment-based approach to development. Case studies in this book (Chapters 2 and 3) have highlighted examples whereby communities have engaged in dialogue with government and this helped to build their confidence, has led to greater interest in and motivation to engage with political processes, and has stimulated the formation of alliances to ensure a greater scale of impact. Despite these motives, it is important to ensure that engagement and influence is credible and builds confidence of government that citizens are worth listening to, while also demonstrating to citizens that dialogue is worthwhile and leads to progress.

Harnessing the power of endogenous actors

An agent that tries to influence policy has great influence over their own activities and outputs and less influence as they move along the results chain, towards outcomes and eventual impacts. Conversely, the further along the results chain, the greater the influence of other actors, and the wider the range of actors that are having an influence. What the different chapters in this volume have illustrated is that there is a variety of ways that each institution can harness the influence of endogenous actors – of other interest groups – in working towards a common goal, and in this way exert greater influence. This is the basis of alliance building yet creating such partnerships appears to be something that is frequently *ad hoc*. A better understanding of the strengths and weaknesses of different institutions or actors that are targeting shared goals is needed to ensure more effective partnerships for influencing policy.

As the "general theory of change" (Figure 1.4) illustrates, many policy interventions start by generating knowledge and awareness of important issues, and may perhaps contribute to strengthening the ability of different actors to engage more effectively in policy dialogue. Some interventions may actively create opportunities for this dialogue through creation of fora, networks and consultations. Other interventions work directly with government, in an advisory capacity for the creation of new or modified policies, or in the implementation of policies. Nevertheless, the long-term goals of "sustainable development, and improved livelihoods and biodiversity conservation" are only likely to be achieved as an outcome of many interventions by many different actors.

Building alliances is therefore important not only for influencing the development of formal and informal government policies, but also for generating momentum for improved public investment. In other words, alliances can play a role not only in formulating government's position and guidelines on an issue, but also in improving and sustaining best practice in government's operations (Holmén, 2002). There are many examples of non-governmental actors working in partnership with government agencies, particularly at local level, to jointly identify new ways of working with communities, with a view to institutionalizing new practices at both local and eventually national levels. By taking a broad view of policy, as this volume does, such practice-based partnerships can be an important part of policy engagement and in many cases may be more effective than lobbying or advocacy: a form of technical advisory input to government.

Measuring influence

Working in alliance to influence long-term multi-stakeholder processes, raises the challenge of identifying and attributing influence to different parties or actions. An elaborate web of actions, influence and associated assumptions can be drawn up, but this is very challenging to monitor, as examples in this book have illustrated. Monitoring so-called "boundary partners" to examine the extent to which they have changed their attitudes and practices is particularly difficult. Often the manifestation of these changes may not be felt for some time after the action. Alternatively, the desired change in behaviour or attitude on one issue may be constrained by other factors requiring further action.

Changing the attitudes of a number of influential actors on a number of relevant issues takes time and patience: something that many donors lack in their drive for short-term accountability. However there are occasions when profound changes can come about quickly, producing dramatic results, but often only as the end product of years of lobbying and influence. Given the importance of such eventual changes, it appears that greater investment is needed in improving the effectiveness of monitoring outcomes and tracking their long-term influence on programme goals.

As stated in Chapter 1, it is assumed that "with a conducive policy environment, the management of natural resources will improve, investments will be greater and more sustainable, and disincentives will be relaxed". This and other major assumptions need to be routinely checked for validity. Nevertheless, the challenge of demonstrating and attributing impact on the lives of the rural poor should not act as a deterrent to policy work if there is reasonable confidence that such work can have a significant impact at some point in the future.

Key steps for strategizing policy influence

Contributions to this book have been developed from a wide array of policy interventions that have been conducted with varying degrees of foresight and planning. From lessons learned through both success and failure, a number of clear lessons emerge about engaging strategically in influencing policy. There are those people for whom engaging in policy dialogue and influencing decision-makers comes very naturally, but for many actors, particularly those who are poorly connected to government, a clearer framework for understanding the necessary steps can be a useful planning tool. Figure 9.1 illustrates the key steps in strategically planning policy influence, which are discussed subsequently.

A vital first step is a thorough diagnosis of the current situation. Policy, in its vaguest sense, is often identified as something to change without exploring deeply which policy (or absent policy) is really holding up progress. Frequently policy messages are developed that are imprecise and fail to consider the difference between policy documents and policy implementation. Many good practices have been developed in the absence of supporting policies, and there are plenty of examples of policies being developed to catch up with, and perhaps regulate or standardize, good practices. An initial diagnosis of which policies and legal frameworks already exist and how they influence positively or negatively a desired outcome is crucial if policy messages are to be targeted and actionable.

Once a thorough analysis of the situation or context has been developed, it is easier to identify targets of policy influence. There are many actors who influence government decision-making at different levels and it is important to consider which ones are the most relevant targets for addressing a particular policy change or adopting a new concept. Policy targets may be government technical advisors, particularly if change is needed in the mechanisms for implementation of an existing

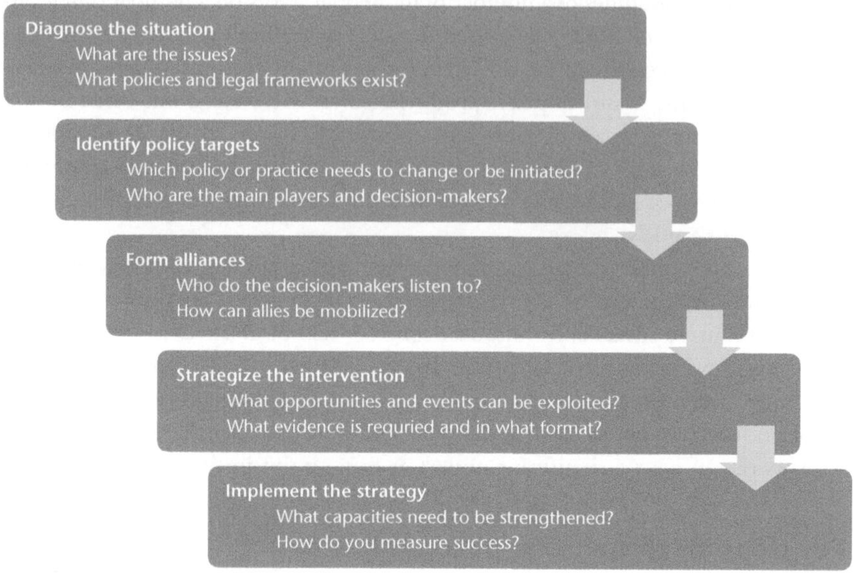

FIGURE 9.1 Strategically planning policy influence

policy, or they may be more political leaders, particularly if the challenge is one of allocating scarce government resources or influencing a political agenda or a policy goal. Successful influencing of policy often hinges on cooperation with a key champion within government, who is connected to decision-making structures and can use new arguments to influence change from within. Targets may also be actors outside government whose support is needed through alliance building but who are uninformed of the challenges or opportunities. This can include Civil Society in the broadest sense, and policy dialogue may target communities, community leaders and Civil Society Organizations to raise key concerns through appropriate channels, such as local development fora of through parliamentary opportunities.

Once the desired policy change and the main players and decision-makers are known it is usually necessary to form suitable alliances and to plan strategically how to engage in policy dialogue. A broad-based alliance will enable a group of actors with different skills, credentials and resources to collaborate in different ways, for example, with different target audiences, and this multi-pronged approach will give an alliance greater effect at different levels of a decision-making process. Within these alliances some actors may play a role of building capacity of others, or may play a convening role, thus enabling other allies to engage in aspects of dialogue that are more suited to their credibility or legitimacy.

Implementation of the strategy depends on the types of opportunities that have been identified, the nature of the evidence or other arguments that are appropriate, and the most suitable means of communication. The role of communication in the policy strategy is paramount, and a broad interpretation of communication is important, as outlined in Chapter 8, to ensure that a wider range of innovative opportunities is considered. For example, in cases of high complexity, participatory approaches can be considered as a means of communication, in which decision-makers participate in action-research with communities and other agencies with a view to identifying policy options. The policy influencing strategy may revolve around exploiting opportunities at public events or policy fora, or may include providing more direct advice or lobbying to decision-makers.

Diagnosing the situation

A critical first step for engaging more effectively in influencing policy is to improve understanding of both the political and the policy context. Although there are many different ways to influence decision-making, all actors can benefit from understanding the opportunities and constraints created by the prevailing political and policy context, as well as the underlying social context. For example, policies of decentralization may offer a degree of autonomy at local level through which change can be instigated, while prevailing attitudes and misconceptions may impede progress locally and might need greater leadership for change at a national level.

The RAPID framework developed by the Overseas Development Institute (ODI) provides a tool for understanding and analyzing the overall context within which policy change needs to take place (Start and Hovland, 2004). This framework addresses the external environment, the key actors and their motivations and influence in decision-making processes. The framework explores the political context and the interest and opportunities for change. The framework also addresses the existence and use of evidence and the links that enable evidence to be brought to the attention of decision-makers (Figure 9.2) (Young, 2005). Analysis of the Political Context is discussed immediately below, whereas the issues of External influences, Evidence and Allies are dealt with in subsequent sections.

The political context

Although the political context may often be taken for granted, or may perhaps be overlooked as being too obvious to policy actors, a thorough analysis may be helpful in identifying key opportunities to influence decision-making. The political context can heavily influence research for policy and the extent of civil and political freedom in a country influences the degree to which researchers can actively inform and advise decision-makers. Where opportunities are few, appropriate partnerships with government may create opportunities for dialogue, for example, through partnership in piloting good practices on the ground. Where

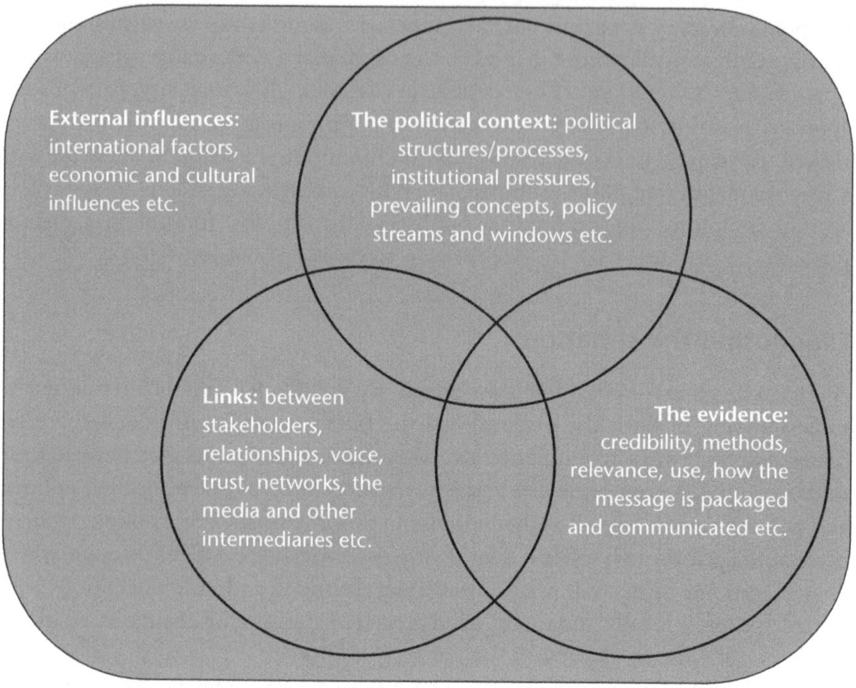

FIGURE 9.2 The RAPID Framework for Assessing Research–Policy Links (Adapted from Start and Hovland, 2004)

one country may be less open to dialogue than its neighbour, regional intergovernmental bodies can play a role in convening dialogue, and exposing decision-makers from one country to the voices and opinions of citizens in the other. Furthermore, the influence of an apparently unrelated policy or government agency may be crucial to addressing the agency that is ultimately responsible.

External influences

The RAPID framework highlights the influence of external factors on research and policy interactions, including the influence of international politics and processes and also the influence of donor policies and research funds. Donor support for Civil Society, in parallel with evolving political freedoms in some countries, has significant influence on the opportunity for dialogue with government. Similarly donor support for research can compromise credibility and raise questions over who is setting the research agenda. Research on development issues is also sometimes conducted in industrialized nations, which can pose challenges of access, relevance and perceived legitimacy.

Analysis of the external environment should explore the roles of different international actors, including the role of Sub-Regional Intergovernmental Institutions, and the interplay between these institutions. For example, the Multilateral Environmental Agreements (e.g. UNCBD, UNCCD, UNFCCC) increasingly engage with subregional bodies to promote environmental actions at the international or transboundary level. However, a critical analysis is needed of the real influence that these institutions have and the forces that they are in turn sensitive to. At the international level, it can also be prudent to explore ongoing economic, political and social processes and trends, and any foreseeable exogenous shocks and trends that could potentially affect the policy process.

Understanding policy processes

Policy processes need to be understood as both national and local level decision-making processes, which can be formal or informal. Questions at the local level may be similar to the national level, although questions around key opportunities and stakeholders will need to be addressed differently. At local level there may be more scope for influence through the influence of community leaders, or conversely there may sometimes be fewer opportunities for real dialogue. This depends not only on the country, but also the location within a country and the degree to which a population is represented by and has good relations with government staff.

Court *et al.* (2006) raise a number of important issues that should be analyzed to improve understanding of policy processes, which summarize the points raised in this section:

1. The macro political context: political freedom; drivers of change; role and vibrancy of Civil Society.
2. Specific policy context: politics of policy agenda setting and formulation, demand for change.
3. Policy implementation: transparency or accountability of bureaucracy; capacity to implement.
4. Timing: key moments and opportunities in the policy change process.
5. The way policy-makers think: clarity of policy objectives; openness to different types of evidence; capacity to process information.

Identify policy targets

One of the principle outputs of the diagnosis detailed in the previous section is the identification of policy targets. When the political and the policy context are better understood, it is possible to define more precisely which policy or practice, and at which level of government, needs to be changed. It is also possible to identify the main actors and decision-makers that need to be influenced (or allied with).

A number of key questions should be posed and answered by policy actors to plan effectively to influence policy:

- Which decision within government do they want to influence?
- How important is such a decision and what is the magnitude of change?
- What does the decision-making process consists of?
- Who plays what role?
- Which factors are most likely to influence the decision?
- Who has the skills to do this effectively?
- What partnerships are needed to improve effectiveness?

The subsequent sections in this chapter explore these questions in greater detail, and the current section looks at the initial diagnosis that should be carried out to understand the existing situation and to inform strategic planning.

An important lesson from a number of the chapters in this book is that influencing government decision-making does not automatically require engagement in national level dialogue over the formulation of policy, or the setting of policy goals and agendas. Many practical successes have been achieved through supporting communities to engage with government at district level and by promoting meaningful participation between government and communities. Partnering with government staff at a local level has provided a fruitful means of getting institutional endorsement while simultaneously building capacity for implementing a new approach or practice. Civil Society Organizations sometimes engage in such partnerships on the ground without recognizing that they are influencing institutional attitudes and judgements, and thereby inadvertently influencing decision-making within those institutions. A greater awareness of the changes that they are contributing to, and a better mechanism for tracking change would be instrumental in capitalizing on progress.

Form alliances

When policy targets have been identified, it becomes easier to assess the type of actors who will be listened to in policy dialogue, and the kind of alliances that need to be built to have most immediate impact. This raises questions of how such alliances can be developed, mobilized and coordinated, and the benefits and opportunities that such alliances can bring.

Alliances to address shared interests or to uphold shared values can take many different forms. Explicit Policy Networks can be created, ranging from highly formalized, sometimes exclusionary, "policy communities" of actors from both inside and outside government, to more loosely structured "issue networks". Public Policy Networks may also be created to span the three sectors of government, business and Civil Society. Civil Society Alliances may also include trade unions, religious groups and grassroots community networks. These alliances may collaborate with governmental actors in the policy process or may act independently. Knowledge-based networks and communities of practice may also be developed, usually oriented towards a theme or an issue and comprising experts and professional bodies (Perkin and Court, 2005).

Formation of alliances and networks is a two-way process, which can facilitate focused information exchange, promote dialogue, make effective use of scarce resources, allow for a collaborative approach on issues of mutual interest, and can strengthen a particular position. These alliances can have varying degrees of formality, some representing a loose coalition of organizations with similar goals, while others may be overseen by a secretariat with members allowed to participate through formal processes of admission and selection. Regardless of their specific structure, NGO and CSO networks are becoming increasingly powerful and crucial agents in international issues (Holmén, 2002). In practice, some Civil Society Organizations prefer to work in isolation, and sometimes in direct opposition to government. Formation of alliances may be seen as anathema to organizations whose public image derives from their independence of action. However, for many organizations the failure to work in alliance can compromise their impact in terms of scope, scale and sustainability, and can sometimes compromise their legitimacy and their public image (Court *et al.*, 2006).

Civil Society alliances play an important role in promoting citizen action, addressing common challenges and sector-wide issues and providing opportunities for joint advocacy and campaigning, knowledge sharing and peer learning. The networks that form around these alliances can play an important role in linking policy-makers with other stakeholders and in bridging research and policy development. There is a degree of consensus that, notwithstanding the barriers, CSO networks harbour great potential to improve effectiveness of international development policy (Perkin and Court, 2005).

The process of formalizing alliances and networks can sometimes act as a constraint on their functioning, particularly when the network structure, rather than networking behaviour, is prioritized. Networks that emerge as offspring from donor-funded projects often struggle with cumbersome systems for coordination and management. However, networking behaviour should be a central goal for organizations attempting to influence policy and it should be recognized that social change is a communication-intensive process and is a core role of Civil Society Organizations (CSOs). CSOs play vital roles of promoting cooperation, spreading information, developing "conscientization", empowering communities and negotiating policy. To do this effectively, they need to learn from each other and share information, and learn from (and represent) grassroots and their organizations. For this, alliance building and networking is essential (Holmén, 2002).

The precise nature or form of each alliance will be determined by their internal and external environments and they may fulfil a variety of functions. They may act as filters, to establish what information should be prioritized, they may act as amplifiers that popularize poorly known or poorly understood ideas, and they may act as convenors that bring different parties together. Alliances can also play facilitating and provisioning roles, with one group of members building the capacity or otherwise strengthening the role of other members. The effectiveness of these alliances can be determined by the level of trust, legitimacy and openness, and by the relationship between key stakeholders, experts and champions (Perkin and Court, 2005).

Alliances are usually designed to build on the strengths of different institutions. Institutions may be better placed to play an inside role of advising government, or lobbying through close engagement with government, relying on a high degree of access to government decision-makers. Other institutions play more of an advocacy or activist role, putting pressure from outside government through demonstrations and other forms of influence. Whether putting pressure from within or from outside, actors may rely on evidence or may make their case largely on a value basis, as illustrated in Figure 9.3.

Alliances can help to maintain an appropriate balance between confrontational and cooperative approaches, and between evidence-based dialogue and value-based dialogue. Legitimacy may be political, legal, technical, moral, and can be conferred in different ways depending on whose opinion counts. Legitimacy may come from the quality of the evidence or the reputation of the scientists or

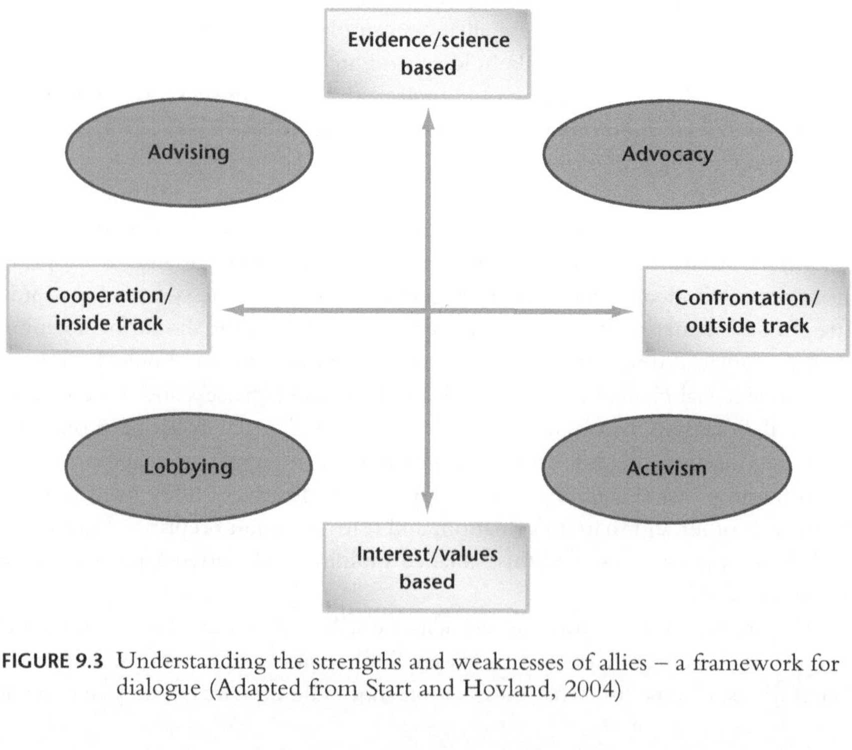

FIGURE 9.3 Understanding the strengths and weaknesses of allies – a framework for dialogue (Adapted from Start and Hovland, 2004)

institutions behind the evidence. Alternatively, legitimacy may stem from the degree and nature of representation of a particular constituency or interest group. Some groups, particularly some Civil Society Organizations or elected representatives, claim legitimacy as representatives of a particular community or population: this is legitimacy based on the weight of popular opinion, and it may sometimes be substituted for evidence. In addition to legitimacy, some actors have particular characteristics or a reputation that makes them more listened-to and influential. Alliances can help to link people with the personality to influence decision-making with other actors who have more persuasive arguments or better-defined recommendations.

Strategize the intervention

Effective policy influence requires each actor, or alliance, to have a clear understanding of the opportunities and events that can be exploited, the most effective ways of exploiting those opportunities, the evidence that is most pertinent and the format in which it should be presented. In strategizing an intervention, an understanding of the steps in the policy process (Chapter 1, Figure 1.2) can be useful in identifying different stages and entry point. On the basis of a thorough diagnosis, the policy influencer can consider whether it is necessary, or possible, to influence the policy agenda, the formulation of policy or the taking of other decisions, the implementation of policy or its evaluation.

The capacity to gather and present the best evidence can be a constraint in policy dialogue, and has frequently been raised as a challenge by Civil Society Actors (Court *et al.*, 2006). Within an alliance, some actors may play a stronger role either in the development of evidence and arguments or in building the capacity of other members to do this. Often the strongest researchers and the best advocates are from different organizations, and the strength of an alliance may lie in its ability to combine scientific credibility with representational legitimacy and individual persuasiveness.

Access to the policy process is frequently a stumbling block, although one that is increasingly being overcome in the policy formulation stage as some governments become more accountable and consultative. Where the political environment is not conducive to engagement by non-governmental actors, campaigns may be necessary, or strategic means of engagement might be pursued, such as hands-on influence of government through government-NGO partnership in particular pilot projects. It is also important to know the key decision-makers and to build up a relationship with them and also to identify champions within relevant government institutions. Through close and sustained dialogue with key actors it is possible to break down barriers, identify sticking points, and reach compromise. It is also possible to identify actors who are implacably opposed to a particular course of action and strategize ways of tackling them (Chapter 8).

Understanding both the policy process and the political environment can help to avoid wasting energy on making the wrong arguments, in the wrong fora, or at the wrong time. Timing is critical, but can be particularly challenging if policy engagement is donor-dependent, which can reduce spontaneity and reactivity. Again, alliances can be helpful here in that those actors with resources at their disposal at an opportune moment can collaborate with other actors that are either better positioned for policy dialogue or have greater capacity.

Assembling relevant and credible evidence can make a significant difference in policy dialogue. Credibility of evidence is often much more about the long-term reputation of the institution or researcher producing it than the credentials of a specific piece of work (JBIC Institute, 2006). The relevance of the work can also depend on the practicality and applicability of the research. Piloting research, and doing so in partnership with the institution that is the target of policy influence, can go a long way towards ensuring that research is noticed and taken seriously. In addition to relevance and credibility, the acceptability of evidence can depend on a number of factors: on its availability, its accuracy, its objectivity and its "generalizability" (Court *et al.*, 2006).

Not all actors in policy dialogue rely on, or even value, evidence, and this may particularly apply to value-based and interest-based advocates. These actors may be important members of an alliance, and may be critical in getting noticed in the first instance. For example, a community group may be instrumental in opening up avenues for dialogue with local government through purely political demands. Nevertheless, evidence should not be shunned entirely, even by value-based advocates, since evidence not only strengthens arguments and credibility, but may also help to guard against poorly informed policy advice that would ultimately fail in its intentions.

It should be noted that evidence and research can influence policy both directly and indirectly. The indirect influence comes through introducing new terms and concepts to the popular discourse and shifting public debates. Evidence can influence public opinion, cultural norms, and political contestation. Evidence can also be used to facilitate the implementation of policy, to influence practice and behaviours on the ground, and for monitoring the outcomes or impacts of policy on the environment or on sustainable development (Young, 2005).

Implement the strategy

Chapters within this book give a variety of examples of ways that policy interventions can be implemented, depending on what policy or decision is being influenced, who the target is, and who is engaging in the process. In its broadest sense, policy is influenced in so many ways that the influence can often be inadvertent, as for example, where government advisors visit a community initiative to give advice but take new ideas away with them, or where a government department partners with an NGO in project implementation. When such

Examples of policy influence outlined in this book

- Influencing local planning and budgeting processes by building confidence of local communities and bringing them into dialogue with local government staff.
- Influencing planning on specific issues at local and national level by linking communities together to build solidarity-based networks.
- Influencing political leadership in setting a policy agenda by strengthening the role of community leaders.
- Influencing a policy agenda by exposing parliamentarians to new ideas and realities in their constituency.
- Influencing policy implementation and budgetary support by partnering with government technical advisors in delivery of innovative development approaches.
- Setting regional and international standards and discourse by partnering with Intergovernmental Bodies and via them their member states.
- Linking local communities with scientists to make research more responsive to local needs, more pertinent to emerging policy debates, and better connected to the end users of policy arguments.

"accidents" are recognized they can be carefully managed to exploit the opportunity for informal and low-level communication and to feed new ideas into government debates, even when those debates are held informally or behind closed doors.

In developing a policy influencing strategy, it may be necessary to consult a wide range of stakeholders to secure their support and buy-in. Indeed, the process of consultation may be at the heart of the strategy and give the policy messages the necessary legitimacy and weight to have a real impact. Effective multi-stakeholder processes need to be skilfully facilitated to be consultative, inclusive, transparent and credible and clearly aimed at an explicit decision. A diverse selection of participants should be included who are there willingly, and who are given an opportunity to speak (Tyler, 2006). These processes can be seen as a social learning process, and as Chapter 2 has illustrated, they can generate enthusiasm for dialogue that can go beyond the immediate issue being addressed.

Communication in its broadest sense is crucial to influence policy, and as we have seen it can be more influential than evidence in influencing policy. Chapter 8 has illustrated the breadth of communication tools that can be used, and in particular some of the less obvious forms of communication that could be of particular relevance to actors, such as Civil Society Organizations, that have poor access to government decision-making. To be effective, communication needs to be tailored to the receptivity of the audience to a given message, the sort of language that will influence them, and the profile of the messengers. The importance of getting communication right is often overlooked, yet effective communication is demanding and requires particular skills and aptitudes.

Although research may often fail to have direct influence on specific policies, the production of research may still exert a powerful indirect influence through introducing new terms and shaping the policy discourse, or by changing public opinion or debates (Weiss, 1977; Lindquist, 2003). These indirect impacts should not be trivialized even though they can be hard to measure or attribute and thorough evaluation can help to draw attention to the accidental side effects of implementing a policy strategy.

Conclusion

The framework and experiences outlined in this book contribute to the current discourse on influencing policy in two ways: broadening understanding of decision-making processes to enable a more strategic approach to dialogue and engagement; and providing practical examples of a wide array of approaches that have successfully led to changes in the way government does business, at national or local level.

The case studies in this volume have not all presented perfect examples of deploying the strategic planning framework outlined in this final chapter. In each example the implementer has faced constraints that have provided important lessons in strategic planning. The examples are quite typical of the work carried out by different actors, particularly NGOs, in the field, and the lessons are similarly relevant to a range of actors. It is evident that policy influence by NGOs is often poorly informed and greater attention should be paid to understanding what needs to be changed, whether it is attitudes, behaviours, procedures or rules. The initial diagnosis is vital to ensure that policy input is directed to the relevant audiences, provides the correct advice, and is delivered in the most effective way.

The experiences presented in this book illustrate the political nature of policy dialogue and highlight ways that different actors with different degrees of impartiality or neutrality can collaborate to achieve a common goal. The chapters have also raised questions over the impartiality of some actors whose neutrality may be compromised depending on where their funding comes from. The inherently political nature of policy dialogue often demands that those engaged in policy dialogue choose sides, even when the role is simply the impartial provision of scientific evidence. By supporting a marginalized group to wield evidence in policy debates, for example, is to align with one stakeholder and empower them in dialogue with other stakeholders. Understanding this will help in harnessing the different political as well as analytical and communication skills of different allies for effective policy dialogue. Those parties with the passion and moral legitimacy to act, but who lack the evidence or the technical skills, may often play the most important role in influencing policy.

This book has shown that there are many ways to influence decision-making at different levels of government, and such influence can have a positive impact on both livelihoods and conservation. It is important to continuously learn how such impact has been achieved in order to improve on it, despite the challenges that lie in convincingly attributing influence. To be effective however, much deeper insight is necessary to really understand the existing policy environment, to identify the most strategic entry points, and to identify allies that have complementary skills and attributes. There is a great deal that actors can achieve to influence government decision-making that lies outside the traditional view of "policy". By influencing the practice of government it is often possible to contribute practically to influencing more formal policy processes. By adopting a broad interpretation of what we mean by policy it is possible to harness the strengths of a wide array of stakeholders in order to bring influence to bear on multiple decision-makers. Working with multiple actors in this way has the added advantage of enabling and empowering a wider range of stakeholders, who will play a critical role in sustaining pressure on decision-makers and ensuring that positive decisions of government are not quickly reversed as soon as pressure is relaxed.

NOTES

1 Conservation as a livelihood strategy: linking policy with practice

1 www.iucn.org

2 Enabling communities to mobilize government support

1 Martin Luther King, Jr in his Nobel Prize Acceptance Speech (10 December 1964).
2 For purposes of this chapter, communities will be defined as groups of people that may or may not be spatially connected, but who share common interests, concerns or identities. They could be local, national or international, with specific or broad interests.
3 Decentralized governance seeks to accelerate the transfer of decision-making from central government to local institutions with the aim of enhancing efficiency, equity and justice in the management and use of local resources to support local development.
4 Capacity building (including civic education and management skills) allows local institutions and communities to perform their assigned responsibilities effectively and efficiently.
5 The Benet are mountain living communities whose livelihoods have been directly linked with the Mt Elgon forest landscape for hundreds of years. Until the early 1980s, the Benet were indigenous forest dwellers settled within the forest. They had agricultural gardens, grazed their livestock, collected medicines, gathered food, and hunted game meat – all within the forest. This however ended in 1983 when Government decided to evict and resettle them outside the forest after it had declared Mt Elgon a National Forest Reserve and subsequently a PA in the 1993.

3 Mobilizing local expertise through networking and empowerment

1 http://iucn.org/about/union/secretariat/offices/esaro/what_we_do/drylands/idrc/influencing_policy/

4 Who represents practitioners in the policy process?

1 Kenya's Parliamentary System will change with implementation of a new constitution passed in August 2010. This constitution proposes a bicameral system with both National Assembly and Senate.
2 Articles 9, 54 and 62 of the Rwandan Constitution.
3 See Burundi Senate Website www.senat.bi/spip.php?article383
4 MINALOC (Ministry of Local Government): www.minaloc.gov.rw/
5 Local Government Systems in Tanzania: in www.tampere.fi/tiedostot/5nCY6QHaV/kuntajarjestelma_tansania_.pdf
7 The Batwa inhabit parts of Southern Uganda, Eastern DRC, Rwanda and Burundi while the Hutu and Tutsi are the two tribes found in both Rwanda and Burundi.
8 www.cdf.gov.rw/ubudehe/UBUDEHEDOCS/aboutubu/rwanda-nprp.pdf
9 This is the lowest administrative unit in Rwanda – see also the section on local governance system in Rwanda (p. 58).
10 *An Inconvenient Truth* is a 2006 documentary film directed by Davis Guggenheim about former United States Vice President Al Gore's campaign to educate citizens about global warming.
11 Hon. Cyrus Muriuki Rutere, MP Imenti North, Kenya 2009.
12 The Council of Ministers is composed of the Ministers of Foreign Affairs and one other focal minister designated by each member state meets at least twice a year.
13 Lake Victoria Basin Commission (LVBC) is a specialized institution of the EAC that is responsible for coordinating the sustainable development agenda of the Lake Victoria Basin. EAC and IUCN signed a Project Implementation agreement for MERECP with LVBC playing the supervisory role. LVBC is accountable to EAC and EALA in terms of audit hence the visit by PAC to establish efficiency on the use of resources and impacts of the Programme.
14 Taken from a short story written in 2007 entitled Harnessing cultural diversity to enhance Lembus forests conservation in Kenya (IUCN, unpublished).

6 Influence without borders: regional approaches to environmental policy dialogue

1 www.unpei.org/
2 www.iucn.org/about/union/secretariat/offices/esaro/what_we_do/drylands/idrc/
3 www.nema.go.ke/index.php?option=com_content&task=view&id=60&Itemid=69
4 www.eac.int
5 www.sadc.int/index/browse/
6 www.nilebasin.org/newsite/index.php
7 http://igad.int
8 www.lvbcom.org/index
9 www.nilebasin.org/newsite/index.php?option
10 www.iucn.org/wisp
11 www.iucn.org/about/union/secretariat/offices/esaro/what_we_do/drylands/idrc/influencing_policy/

7 Shaping policies: science–policy interface in natural resources management

1 With special thanks to Abdalla Said Shah for his contribution to this chapter.
2 www.gmoafrica.org/2006/10/kenya-approves-national-policy-on.html
3 www.biosafetyafrica.org.za/index.php/2007021155/Uganda-GMO-Legislation/menu-id-100025.html

4 www.biosafetyafrica.org.za/index.php/2007021156/Tanzania-GMO-Legislation/menu-id-100025.html

8 Bringing conservation science to life: the role of communication to support policy influencing processes

1 The authors wish to thank CEC members Grace Mwaura and Umar Kawu for their useful comments, suggestions and examples.
2 In particular the authors wish to thank Jonathan Davies, Barbara Nakangu and Abdalla Shah for the examples they provided.
3 Communication, Education and Public Awareness (CEPA), a toolkit for National Focal Points and NBSAP Coordinators, CBD and IUCN, Montreal 2007
4 This diagram is an adaptation of a diagram by Les Robinson (www.enablingchange.com.au/What's_best.pdf).

BIBLIOGRAPHY

Aboud, A., de Leeuw, J., Said, M., Notebaert, A., Roba, G.M., Manzano, P., Davies, J., Omondi, S., Odhiambo, M.O. and Ogeto, M. 2012. *Natural Resource Management and Biodiversity Conservation in the Drylands of Eastern and Central Africa*. ASARECA: Entebbe.

Adams, W.M. and Hulme, D. 2001. Conservation and Communities: changing narratives, policies and practices in African conservation. In Hulme, D. and Murphree, M. (ed.) *African Wildlife and Livelihoods: the promise and performance of community conservation*, pp. 9–23. James Currey: London.

Adams, W.M. and Hutton, J. 2007. "People, Parks and Poverty: political ecology and biodiversity conservation," *Conservation and Society*, 5 (2): 147–183.

Adams, W.M. and Jeanrenaud, S.J. 2008. *Transition to Sustainability: towards a humane and diverse world*. International Union for Conservation of Nature (IUCN). Gland (Switzerland): IUCN.

Agrawal, A. 1995. "Indigenous and scientific knowledge: some critical comments," *IK*, 3 (3): 3–6.

Agrawala, S. 1998. "Context and early origins of the intergovernmental panel on climate change," *Climatic Change*, 39, 605–620.

Alexis, M. and Chang-Sam, A. 2006. *Fisheries Industry of the Seychelles: at a crossroad*. Central Bank of Seychelles, First Quarterly Review 2006, Vol. XXIV, No. 1. www.cbs.sc/acro/PapersforWebiste/FisheriesIndustry.pdf.

AU. 2005. *The Role of Regional and Intergovernmental Organizations in Promoting Gender Equality*, Commission on the Status of Women, ADVANCE UNEDITED VERSION Panel VII. www.un.org/women's/watch/daw/review/documents.

Bangura, Y. 2004. *Technocratic Policy-making and Democratic Accountability*. Technocratic styles of policy-making pose a threat to democracy, because decisions are not subjected to the public scrutiny and dynamics of bargaining that are at the heart of democratic politics. Policy Brief 3 (online). United Nations Institute for Social Development, Geneva: Switzerland. www.unrisd.org/publications/rpb3e.

Bangura, Y. 1994. "Intellectuals, Economic Reform and Social Change: Constraints and opportunities in the formation of a Nigerian technocracy," CODESRIA Monograph Series I/94.

Barkan, J.D. and Matangi, F. 2009. "Kenya's Tortuous Path to Successful Legislative Development." In Barkan, J.D. 2009. *Legislative Power in Emerging African Democracies*. pp. 33-72, Boulder, CO.: Lynne Rienner Publishers.

Barkenbus, J. 1998. *Expertise and the Policy Cycle. Energy, Environment, and Resources Center.* Knoxville, TN: The University of Tennessee.

Baron, N. 2010. *Escape from the Ivory Tower: a guide to making your science matter.* Washington, DC: Island Press.

Bartle, P. 2000. *Elements of Community Strength*. www.scn.org/cmp/empo.htm (accessed on 9 December 2010).

Berkes, F. 2009. "Evolution of co-management: role of knowledge generation, bridging organizations and social learning," *Journal of Environmental Management*, 90, 1692–1702.

Berkes, F., Colding J. and Folke, C. 2000. "Rediscovery of Traditional Ecological Knowledge as Adaptive Management," *Ecological Applications*, 10, 1251–1262.

Berkes, F., Colding, J. and Folke C. (eds) 2003. *Navigating Social-Ecological Systems: building resilience for complexity and change.* Cambridge: Cambridge University Press.

Best, A., Giljum, S., Simmons, C., Blobel, D., Lewis, K., Hammer, M., Cavalieri, S., Lutter, S. and Maguire, C. 2008. *Potential of the Ecological Footprint for Monitoring Environmental Impacts From Natural Resource Use: analysis of the potential of the ecological footprint and related assessment tools for use in the EU's thematic strategy on the sustainable use of natural resources.* Report to the European Commission, DG Environment. http://ec.europa.eu/environment/natres/pdf/footprint.pdf.

Bishop-Sambrook, C. 2005. *Contribution of Farm Power to Smallholder Livelihoods in Sub-Saharan Africa.* Agricultural and Food Engineering Technical Report 2. Food and Agriculture Organization of the United Nations, Rome, 2005. www.fao.org/docrep/009/a0229e/a0229e00.htm#Contents.

Bobrow, D. and Whalen, J. 2002. "Community Knowledge Sharing in Practice: the Eureka story," *Reflections*, 4 (2), 47–59.

Braithwaite, R.L., Blanchi, C. and Taylor, S.E. 1994. "Ethnographic approach to community organization and health empowerment," *Health Education Quarterly*, 21, (3): 407–416.

Briggs, J. 2005. "The use of indigenous knowledge in development: problems and challenges," *Progress in Development Studies*, 5: 99–114.

Brint, S. 1994. *In an Age of Experts: the changing role of professionals in politics and public life.* Princeton, NJ: Princeton University Press.

Brosio, G. 2000. *Decentralisation in Africa*. www.imf.org/external/pubs/ft/seminar/2000/fiscal/brosio.pdf.

Brown, A.L. and Tomerini, D. 1998. *Environmental Mainstreaming in Developing Countries.* Urban Research Program, Griffith School of Environment, Griffith University: Nathan, Australia.

Brown, J.R and Mcleod, N.D. 1996. "Integrating ecology into natural resources management policy," *Journal of Environmental Management*. 20 (3): 289–296.

Bruckmeier, K. and Tovey, H. 2009. *Rural Sustainable Development in the Knowledge Society.* Aldershot: Ashgate.

Carden, F. 2009. *Knowledge to Policy: making the most of development research.* Ottawa, and Sage, New Delhi: International Development Research Centre.

CDC. 1997. *Principles of Community Engagement*. CDC/ATSDR Committee on Community Engagement. www.cdc.gov/phppo/pce/.

Centeno, M.A. 1998. *The Politics of Expertise in Latin America.* London and New York: Macmillan Press.

Chavis, D.M. and Wandersman, A. 1990. "Sense of community in the urban environment: a catalyst for participation and community development," *American Journal of Community Psychology*.

Chhetri, P.B., Barrow, E.G.C. and Muhweezi, E. (eds) 2004. *Securing Protected Area Integrity and Rural People's Livelihoods: lessons from twelve years of the Kibale and Semliki Conservation and Development Project*. Nairobi: IUCN. 168 pp.

Clay, E.J. and Schaffer, B.B. (eds) 1986. *Room for Manoeuvre: an explanation of public policy in agriculture and rural development*. London: Heinemann.

Coleman, J.C. 1988. "Social capital in the creation of human capital," *American Journal of Sociology*, 94: S95–S120.

Conaghan, C. 1996. "A deficit of democratic authenticity: political linkage and the public in Andean politics," *Studies in Comparative International Development*, 31 (3) Fall: 32–55.

Corrales, J. 1997. "Why Argentines Followed Cavallo: a technopol between democracy and economic reform." In Domínguez, J.I. (ed.), *Technopols: freeing politics and markets in Latin America*. University Park, PA: Penn State University Press.

Corrales, J. 2002. *Presidents Without Parties: economic reforms in Argentina and Venezuela*. University Park, PA: Pennsylvania State University Press.

Costanza, R., d'Arge, R., de Groot, R., Farber, S., Grasso, M., Hannon, B., Limburg, K., Shanid, N., O'Neill, R.V., Paruelo, J., Raskin, R. G., Sutton, P. and van den Belt, M. 1997. "The value of the world's ecosystem services and natural capital," *Nature*, 387, 253–260.

Court, J., Mendizabal, E., Osborne, D. and Young, J. 2006. *Policy Engagement: how Civil Society can be more effective*. Overseas Development Institute: www.odi.org.uk/resources/download/160.pdf.

Dalal-Clayton, B. and Bass, S. 2009. *The Challenges of Environmental Mainstreaming: experience of integrating environment into development institutions and decisions*. London: International Institute for Environment and Development.

Davies, J.C. and Mazurek, J. 1998. *Pollution Control in the United States: evaluating the system*. Washington, DC: Resources for the Future.

Davies, J. and Hatfield, R. 2008. "The economics of mobile pastoralism: a global summary," *Nomadic Peoples*, 11: 91–116.

Davies, J., Niamir-Fuller, M., Kerven, K. and Bauer, K. 2010. "Extensive livestock production in transition: the future of sustainable pastoralism." In Steinfeld, H., Mooney, H.A., Schneider, F., and Neville, L.E. (eds) *Livestock in a Changing Landscape, Volume 1, Drivers, Consequences, and Responses*. Washington, DC: Island Press.

Dearing, J.W. and Rogers, E.M. 1996. *Agenda-Setting*. Thousand Oaks, CA: Sage Publications.

EAC. 2011. *EAC portal*. www.eac.int/about-eac.html.

Earl, S., Carden, F. and Smutylo T. 2001. *Outcome Mapping: building learning and reflection into development programs*. Ottawa, Canada: International Development Research Centre (IDRC).

ECA. 2007. Relevance of African Traditional Institutions of Governance Workshop. Addis Ababa, Ethiopia, 17–18 June 2006. UN Economic Commission for Africa.

Eden, S. 1996. "Public participation in environmental policy: considering scientific, counter-scientific and non-scientific contributions," *Public Understanding of Science*, 5: 183–204.

Eden, S. 1998. "Environmental Issues: knowledge, uncertainty and the environment," *Progress in Human Geography*, 22 (3): 425-32.

Emery, A.R. 1996. *The Participation of Indigenous Peoples and Their Knowledge in Environmental Assessment and Development Planning* (draft). Ottawa, Canada: Centre for Traditional Knowledge.

Emorut, S. 2006. "Decentralization and accountability in infrastructure delivery in developing countries," *Economic Journal*, 116 (January): 101–127.

ESRC. 1998. *Strengthening Decision-making for Sustainable Development*. In Report of a Workshop held at Eynsham Hall, Oxford, 15–16 June 1998.

Fawcett, S.B., Paine-Andrews, A., Francisco, V.T., Schultz, J.A., Richter, K.P., Lewis, R.K., Williams, E.L., Harris, K.J., Berkley, J.Y., Fisher, J.L. and Lopez, C.M. 1995. "Using empowerment theory in collaborative partnership for community health and development," *American Journal of Community Psychology*, 23 (5).

Fazey, I., Fazey J., Salisbury, J., Lindenmayer, D. and Dovers, S. 2006. "The nature and role of experiential knowledge for environmental conservation," *Environmental Conservation*, 33: 1–10.

Fisher, R.J., Maginnis, S., Jackson, W.J., Barrow, E. and Jeanrenaud, S. 2005. *Poverty and Conservation: landscapes, people and power*. Switzerland (Gland): IUCN.

Flintan, F. 2008. *Study on Good Practice: Women's empowerment in Pastoral Societies*, Kenya.

Ford, J. and Martinez, D. 2000. "Traditional ecological knowledge, ecosystem science and environmental management," *Ecological Applications*, 10 (5): 1249–1250.

Friedrich Ebert Foundation and Legis Consult. 2003. *Working Structures of Parliament in East Africa*. http://library.fes.de/pdf-files/bueros/kenia/01704content.pdf.

Fujisaka, S. 1997. "Research: help or hindrance to good farmers in high risk systems?," *Agricultural Systems*, 54(2): 137–152.

Gaventa, J. and Robinson, M. 1999. *Influence from Below and Space from Above: non elite action and pro poor policies*. Paper prepared for Poverty Conference, Institute of Development Studies, June 1998 (unpublished), cited in Holmes, T. and Scoones, I. 2000.

Grenier, L. 1998. *Working with Indigenous Knowledge: a guide for researchers*. Ottawa, Canada: International Development Research Centre (IDRC).

Guchteneire, P., Krukkert, I. and Liebenstein, G. 2010. *Best Practicies in Indigenous Knowledge*. Management of Social transformations Programme (MOST) and the Centre for International Research and Advisory Networks (CIRAN). www.unesco.org/most/bpikpub.htm.

Guldin, R.W., Parrotta, J.A and Hellström, E. 2005. *Working Effectively at the Interface of Forest Science and Forest Policy*. Guidance for Scientists and Research Organizations. IUFRO Occasional Paper No. 17.

Habermas, J. 1984. *The Theory of Communicative Action – Reason and the Rationalisation of Society* (Vol. I). Boston, MA: Beacon Press.

Hailler J.-P. 2008. *Status of the Indian Ocean Tuna Tagging Programme – RTTP-IO*. Working Party on Tagging Data Analysis meeting, Victoria, Seychelles, 30 June–4 July, 2008. www.iotc.org/files/proceedings/2008/wptda/IOTC-2008-WPTDA-10.pdf.

Hart, S. (ed.) 2008. *Shared Resources: Issues of Governance*. Gland, Switzerland: IUCN. 261 pp.

Healey, P. 1997. *Collaborative Planning: shaping places in fragmented societies*. London: Macmillan.

Held, I.M. and Soden, B.J. 2000. "Water vapor feedback and global warming," *Annual Review of Energy and the Environment*, 25: 441–475.

Hirji, R. and Davis, R. (eds) 2009. *Environmental Flows in Water Resources Policies, Plans, and Projects*. Case studies. Washington, DC: World Bank.

Hoekstra A.Y. 2009. "Human appropriation of natural capital: a comparison of ecological footprint and water footprint analysis," *Ecological Economics*, 68: 1963–1974.

Holmén, H. 2002. *NGOs, Networking, and Problems of Representation*. Linköping, Sweden: Linköpings University and ICER.

Holmes, T. and Scoones, I. 2000. *Participatory Environmental Policy Processes: experiences from north and south*. IDS Working Paper 113. Brighton, UK: Institute of Development Studies, University of Sussex.

Hossfeld, U. and Olsson L. 2002 "From the Modern Synthesis to Lysenkoism, and Back?,"*Science*, 297, 55–56.

Ingram, H.M. and Mann D.E. 1980. *Why Policies Succeed or Fail*. Beverly Hills, CA: Sage Publications.

Ingram, J. 2008. "Are farmers in England equipped to meet the knowledge challenge of sustainable soil management? An analysis of farmer and advisor views," *Journal of Environmental Management*, 86: 214–228.

IUCN and IGAD. 2007. *Environment and Natural Resources as Core Assets for Wealth Creation Poverty Reduction and Sustainable Development*, IGAD conference Workshop report, www.cmsdata.org.

IUCN. 2007. Second IGAD Directors of Conservation and Economic Planning Conference, 9–11 March, 2010, Imperial Botanic Beach Hotel, Entebbe, Uganda. http://cmsdata.iucn.org/downloads/igad_conference_report_.pdf.

IUCN, 2008. *First IGAD Parliamentarians Tour: policy briefing note*. http://cmsdata.iucn.org/downloads/mptour_pastoralism_policybrief_09.pdf.

IUCN. 2010. Noxious weed threatens the biggest wildlife migration on the planet. SSC Press Release. www.iucn.org/about/work/programmes/species/news_events/?6511/Noxious-weed-threatens-the-biggest-wildlife-migration-on-the-planet.

IUCN/UNEP/WWF, 1991. *Caring for the Earth: a strategy for sustainable living*. Gland, Switzerland: IUCN/UNEP/WWF.

Iwona, S. and de Vries, M.S. 2002. *The Role of Outside Experts in Local Government*. Paper for the 12th NISPAcee Annual Conference: Central and Eastern European Countries Inside and Outside the European Union: Avoiding a New Divide. Vilnius: Lithuania, May 13–15, 2004.

Jacques, P.J., Dunlap, R.E. and Freeman, M. 2008. "The organisation of denial: conservative think tanks and environmental scepticism," *Environmental Politics*, 17, 349–385.

James, S. 1993. "The idea brokers: the impact of think tanks on British government," *Public Administration*, 71 (4) Winter: 491–506.

JBIC Institute. 2006. *How to Influence Policy in the African Energy Sector*. JBICI Discussion Paper No. 13. July 2006. Japan Bank for International Cooperation.

Johnson, E.C. (1996) "How think tanks improve public policy." *Economic Reform Today*, 3: 1996.

Jones H., Jones N., Walker, D. and Walksh, C. 2009. *Strengthening Science–Policy Dialogue in Developing Countries: a priority for climate change adaptation*. Background Note. London: Overseas Development Institute.

Jordan, J. and Jones, P. (1997) "Assessing your company's knowledge management style," *Long Range Planning*, 30: 392–398.

Joshi, L. and Shrestha, P. (no date). *Local Ecological Knowledge for Healthy Agriculture*, SE Asia, Indonesia.

Jutting, J., Kauffmann, C., McDonnell, I., Osterrieder, H., Pinaud, N. and Wegner, L. 2004. *Decentralisation and Poverty in Developing Countries: exploring the impact, OECD Development Centre*, Working Paper No. 236, Paris, France. www.oecd.org/dataoecd/40/19/33648213.pdf.

Kambona, J.J. and Marashi, S.H. 1996. *Process for the Establishment of the Indian Ocean Tuna Commission*. FAO, Rome. www.fao.org/docrep/W1750E/W1750E00.htm.

Kandlikar, M. and Ambuj Sagar, A. (1997) *Climate Change Science and Policy: lessons from India*. ENRP Discussion Paper E-97-08. Cambridge, MA: Kennedy School of Government, Harvard University.

Kauzya, J. M. December 2007. *Political Decentralization in Africa: experiences of Uganda and South Africa*. Accessed at: http://unpan1.un.org/intradoc/groups/public/documents/un/unpan028411.pdf.

Keeley, J. and Scoones, I. 1999. *Understanding Environmental Policy Processes*. IDS Working Paper 89. Brighton, UK: Institute of Development Studies.

Kenya National Land Policy (draft) 2010. National land policy secretariat, Ministry of Lands, Government of Kenya, Nairobi.

Kettl, D. 2009. *The Next Government of the United States: why our institutions fail us and how to fix them*. New York: WW Norton.

Khoo, B.T. 2010. *No Insulation: politics and technocracy's troubled trajectory*. IDE Discussion Paper No. 236. 2010. 05. http://hdl.handle.net/2344/897.

Kigenyi, F., Gondo, P. and Mugabe, J. 2000. *Practice Before Policy: an analysis of policy and institutional changes enabling community involvement in forest management in Eastern and Southern Africa*. Forest and Social Perspective in Conservation No. 10.

Kingdon, J. 1984. *Agendas, Alternatives and Public Policies*. Boston/Toronto, Canada: Little Brown & Company.

Kolk, A. and Levy, D. 2001. "Wind of change: corporate strategy, climate change and oil multinationals," *European Management Journal*, 19, 501–509.

Kundishora, P. 2009. *An Overview of Decentralization and Local Governance in Eastern and Southern Africa*. http://mpra.ub.uni-muenchen.de/15701/.

Lane, S.N. 2001. Constructive comments on D. Massey, "Space-time, 'science' and the relationship between physical geography and human geography." *Transactions of the Institute of British Geographers*, 26: 243–256.

Langill, S. 1999. *Indigenous Knowledge: a resource kit for sustainable development researchers in dryland Africa*. People, Land and Water Program Initiative. Ottawa, Canada: IDRC.

Lapintie, K. 1998. "Analyzing and evaluating argumentation in planning," *Environment and Planning B: Planning and Design*, 25: 187–204.

Lasswell, H. 1977. "The Politics of Prevention," in *Psychopathology and Politics*, Chicago: University of Chicago Press.

Legros, G., Havet, I., Bruce, N. and Bonjour, S. 2009. *The Energy Access Situation in Developing Countries: a review focusing on the least developed countries and sub-Saharan Africa*. http://content.undp.org/go/cms-service/stream/asset/?asset_id=2205620.

Leshner, A. 2002. "Advancing science, serving society," *Science*, 295 (5557): 929, 8 February 2002.

Leverington, F., Hockings, M., Pavese, H., Lemos Costa, K. and Corrau, J. 2008. *Management Effectiveness Evaluation in Protected Areas – a global study*. Supplementary Report No. 1: Overview of approaches and methodologies. TNC, WWF, IUCN-WCPA: University of Queensland, Gatton, Australia. http://cmsdata.iucn.org/downloads/methodologyreportdec08_final.pdf.

Li, C.C. 1987. "Lysenkoism in China," *The Journal of Heredity*, 78: 339–340.

Lindquist, E.A. 2003. *Discerning Policy Influence: framework for a strategic evaluation of IDRC-supported research*. Ottawa, Canada, International Development Research Centre (IDRC).

Long Martello, M. 2004. "Expert advice and desertification policy: past experience and current challenges," *Global Environmental Politics*, 4: 85–106.

Lorenzoni, I. and Hulme M. 2009. "Believing is seeing: laypeople's views of future socioeconomic and climate change in England and in Italy," *Public Understanding of Science*, 18 (4): 383–400.

Lowi, T.J. 1998. "Foreword." In Tatalovich, R. and Daynes B.W. *Moral Controversies in American Politics: cases in social regulatory policy*, 4th edn. New York: Armonk.

Ludwig, D., Hilborn, R. and Walters C. 1993. "Uncertainty, resource exploitation, and conservation: lessons from history," *Science*, 260: 17 + 36.

Lindvall, J. 2006. "The Politics of Purpose: economic policy in Sweden after the Golden Age," *Comparative Politics*, 38: 3.

Lundvall, B. and Johnson, B. 1994. "The learning economy," *Journal of Industry Studies*, 1: 23–42.

LVBC. 2011. *Lake Victoria Basin Commission*. www.lvbcom.org. (Retrieved November 2010.)

McCright, A.M. and Dunlap, R.E. 2003. "Defeating Kyoto: the conservative movement's impact on U.S. climate change policy," *Social Problems*, 50: 348–373.

McGregor, D. 2004.*Traditional Ecological Knowledge and Sustainable Development: towards coexistence*. www.idrc.ca/ggp/ev-64525-201-1-DO_TOPIC.html.

McNeely, J.A. 1993. "Economic incentives for conserving biodiversity: lessons for Africa," *Ambio*, 22: 144–150.

MA. 2005. *Living Beyond Our Means: natural assets and human wellbeing*. Available at www.maweb.org/documents/document.429.aspx.pdf.

Mace, G.M., Cramer, W., Díaz, S., Faith, D.P., Larigauderie, A., Le Prestre, P., Palmer, M., Perrings, C., Scholes, R.J., Walpole, M., Walther, B.A., Watson, J.E.M. and Mooney, H.A. 2010. "Biodiversity targets after 2010," *Current Opinion in Environmental Sustainability*, 2: 1–6.

Mackinson, S. and Nottestad, L. 1998. "Combining local and scientific knowledge," *Reviews in Fish Biology and Fisheries*, 8: 481–490.

Mafabi, P., Kansiime, F., Opio, A. and Busuluwa, H. 2005. "Kinawataka Wetland: a functional ecosystem that could be used as a reference to conserve wetlands in Uganda." In Vymazal, J. (ed.) *Natural and Constructed Wetlands: nutrients, metals and management*. Leiden, Netherlands: Backhuys Publishers.

Mander, J. 1991. *In the Absence of the Sacred: the failure of technology and the survival of the Indian nations*. San Francisco: Sierra Club Books.

Manzano, P., Azcarate, F.M., Peco, B. and Malo, J.E. 2010. "Are ecologists blind to small things? The missed stories on non-tropical seed predation on feces," *Oikos*, 119: 1537–1545.

Maravall, J.M. and Przeworski, A. 2003. *Democracy and the Rule of Law*. Cambridge: Cambridge University Press.

Mathews, Jessica. T. 1996. "Creating an Independent Institute for Policy Research on Global Issues: the World Resources Institute." In Telgarsky, J. and Ueno, M. (eds) *Think Tanks in a Democratic Society: an alternative voice*. Washington, DC: Urban Institute.

Meadows, D. 2008. *Thinking in Systems: a primer*. London: Earthscan.

Mehrabian, A. 1981. *Silent Messages: implicit communication of emotions and attitudes*, 2nd edn. Belmont, CA: Wadsworth.

Mendizabal, E. 2008. *Supporting networks: ten principles*, ODI Opinion 105. London: ODI.

Milani, C.R.S. 2009. *Evidence-based Policy Research: critical review of some international programmes on relationships between social science research and policy-making*. Paris: United Nations Educational, Scientific and Cultural Organisation.

Millennium Ecosystem Assessment. 2005. *Ecosystems and Human Well-being: synthesis*. Washington, DC: Island Press.

Ministry of Water and Livestock Development. 2002. National Water Policy – July 2002, Internet version. The United Republic of Tanzania. www.ewura.go.tz/pdf/National WaterPolicy.pdf.

Moore, C. 2004. "Regional Integration and Regional Governance under the New African Initiatives: A Critical appraisal," Centre for Policy Studies. *Policy: Issues and Actors*, 17 (3). www.oecd.org/data/oecd.

Mortimore, M. with contributions from Anderson, S., Cotula, L., Davies, J., Faccer, K., Hesse, C., Morton, J., Nyangena, W., Skinner, J. and Wolfangel, C. 2009. *Dryland Opportunities: a new paradigm for people, ecosystems and development*. Gland, Switzerland: IUCN; London, UK: IIED; and Nairobi, Kenya: UNDP/DDC. 96 pp. www.iucn.org/dbtw-wpd/edocs/2009-033.pdf.

Moyo, B. 2009. "Indigenous knowledge-based farming practices: a setting for the contestation of modernity, development and progress," *Scottish Geographical Journal*, 125: 353–360.

Mugabe, J. and G.W., Tumushabe 1999. "Environmental Governance: conceptual and emerging issues." In Okoth Ogendo, H.W.O. and Tumushabe, G.W. (eds). *Governing the Environment: political change and natural resource management in Eastern and Southern Africa*. Nairobi, Kenya: African Centre for Technology Studies.

Mugabi, E., 2004. *Uganda's Decentralisation Policy, Legal Framework, Local Government Structure and Service Delivery*. http://unpan1.un.org/intradoc/groups/public/documents/UN/UNPAN029080.pdf.

Mwaipopo, R. 2010. *Building Policy for Supporting Livelihoods Through Conservation: review of bottlenecks to influencing environmental and development policy*. Nairobi, Kenya: IUCN ESARO, pp. iv, 18.

Mzava, E., Mbura, M., Bwindiki, J., Uronu, G., van Ingen, T. and Wells, S. 2007. "Capacity Development." In Wells S., Makoloweka, S. and Samoilys, M. (eds) *Putting Adaptive Management into Practice: collaborative coastal management in Tanga, Tanzania*. IUCN-EARO, Nairobi.

Narayan, D. 1997. *Voices of the Poor: poverty and social capital in Tanzania*. Environmentally and Socially Sustainable Development Studies and Monographs, Series 20. Washington, DC: World Bank.

NBI. 2011. *Nile Basin Initiative*. www.nilebasin.org/newsite/. (Retrieved May 2011.)

Nelkin, D. (ed.) 1992. *Controversy: Politics of Technical Decisions*. London: Sage.

Neumann, R.P. 2004. Nature-State-Territory: towards a critical theorisation of conservation enclosures. In *Liberation Ecologies: environment, development, social movements*, R. Peet and M. Watts (eds), London: Routledge, pp. 195–217.

Niemi, E. and Manyindo, J. 2010. *Economic Importance of Goods and Services Derived from Dry land Ecosystems in the IGAD Region: case studies*. Nairobi: International Union for Conservation of Nature. http://cmsdata.iucn.org/downloads/4615_ful_report_10_0421_2_pdf.

Northouse, P.G. 2010. *Leadership: theory and practice* 5th edn. Thousand Oaks, CA: Sage.

Nyangabyaki, B. 2003. *Decentralisation, Politics and Environment in Uganda*. Washington, DC: World Resources Institute.

Oba, G. 2009. *Harnessing Pastoralists' Indigenous Range Management Knowledge: selection and application of indicators for decision-making for drought resilient livelihood systems in the Horn of Africa*. July, 2009. Nairobi: IUCN/WISP.

O'Donnell, G. 1994. "Delegative democracy," *Journal of Democracy*, 5 (1): 55–69.

Oh, C.H. 1996. *Linking Social Science Information to Policy-making*. Greenwich, CT: Jai Press.

Okidi, J.A. and Guloba, M. 2006. *Decentralisation and Development: emerging issues from Uganda's experience*. Economic Policy Research Centre, Kampala, Uganda. http://ageconsearch.umn.edu/bitstream/93810/2/op31.pdf.

Olan'g, S. 2010. *Tanzania Passes a New Mining Law and Builds Capacity for Informed Policy Debate*. Revenue Watch Institute, 21 May 2010. www.revenuewatch.org/news/tanzania-passes-new-mining-law-and-builds-capacity-informed-policy-debate.

Ole Nasha, W. 2007. *Pastoralism as a Conservation Strategy in Tanzania*. Nairobi, Kenya: IUCN-EARO.

Oleh, P. 2004. "Ruling with decrees: presidential decree making in Russia and Ukraine," *Europe-Asia Studies*, 56 (5) July: 637–660.

Olsen, R. 2009. *Do Not Be Such a Scientist: talking substance in an age of style*. Washington, DC: Island Press.

O'Neil, M. 2009. "Preface." In Carden, F. *Knowledge to Policy: making the most of development research*. International Development Research Centre, Ottawa, and Sage, New Delhi

Ostrom, E. 1990. *Governing the Commons: the evolution of institutions for collective action*. New York: Cambridge University Press.

Ostrom, E. 2005. *Understanding Institutional Diversity*. Princeton, NJ: Princeton University Press.

Perkin, E. and Court, J. August 2005. *Networks and Policy Processes in International Development: a literature review*, Working Paper 252. London: Overseas Development Institute.

Polacheck. T. 2006. "Tuna longline catch rates in the Indian Ocean: did industrial fishing result in a 90% rapid decline in the abundance of large predatory species?," *Marine Policy*, 30: 470–482.

Pollard, A. and Court, J. 2005. *How Civil Society Organisations Use Evidence to Influence Policy Processes: a literature review*. Working Paper 249, July 2005, Overseas Development Institute.

Portney, K. 1988. "Foreward." In Davies, J.C. and Mazurek, J. *Pollution Control in the United States: evaluating the system*. Washington, DC: Resources for the Future.

Putnam, R. 1993. *Making Democracy Work: civic traditions in modern Italy*. Princeton, NJ: Princeton University Press.

Ramsar Convention Secretariat. 2010. "National Wetland Policies: developing and implementing National Wetland Policies." *Ramsar Handbooks for the Wise Use of Wetlands*, 4th edn, vol. 2. Gland, Switzerland: Ramsar Convention Secretariat.

Reed, M., Evely, C., Cundill, G., Fazey, I., Glass, J., Laing, A., Newig, J., Parrish, B., Prell, C., Raymond, C. and Stringer, L. 2010. "What is Social Learning?," *Ecology and Society*, 15 (4). r1. www.ecologyandsociety.org/vol15/iss4/resp1/.

Republic of Kenya. 2005. *Office of the President, Cabinet Office Handbook on Governing Responsibility in Kenya*. www.marsgroupkenya.org/Reports/Government/mar_07/HANDBOOK_ON_GOVERNING_RESPONSIBILITY_IN_KENYA.pdf.

Republic of Uganda. 1995. *National Policy for the Conservation and Management of Wetland Resources*. www.foodnet.cgiar.org/SCRIP/docs&databases/ifpriStudies_UG_nonScrip/pdfs/Nationa_Wetlands_Programme/Uganda%20National%20Wetlands%20Policy%201995.pdf.

Reynolds, F., Stafford, M. and Lambin, F. 2007. "Global desertification: building a science for dryland development," *Science*, 316: 847-851.

Robinson, J. and Shroff, J. 2004. "The fishing sector in Seychelles: an overview, with an emphasis on artisanal fisheries," *Seychelles Medical and Dental Journal* (Special Issue), 7: 52–56.

Rodrigues, A.S.L., Pilgrim, J.D., Lamoreux, J.F., Hoffmann, M. and Brooks, T.M. 2006. "The value of the IUCN Red List for conservation," *Trends in Ecology and Evolution*, 21: 71–76.

Rossi, J. 1997. "Participation run amok: the costs of mass participation for deliberative agency decision-making," *Northwestern University Law Review*, 92 (1): 173–249.

Ruess, R.W. and McNaughton, S.J. 1987. "Grazing and the dynamics of nutrient and energy regulated microbial processes in the Serengeti Grasslands," *Oikos*, 49: 101-110.

Sagasti, F. and Alcalde, G. 1999. *Development Cooperation in a Fractured Global Order: an arduous transition*, IDRC, p. 204.

Sanderson, I. 1999. "Participation and democratic renewal: from 'instrumental' to 'communicative rationality," *Policy and Politics*, 27 (3): 325–341.

Schlee, G. (in press). A comment on the 'Policy framework for pastoralism in Africa' adopted by the African Union in January 2011. *Nomadic Peoples*, 14.

Schneider, A.L. and Ingram, H. 1997. *Policy Design for Democracy*. Lawrence, KS: University of Kansas Press.

Selman, P. 1998. "Local agenda 21: substance or spin," *Journal of Environmental Planning and Management*, 41 (5): 533–553.

Sen, A. 1999. *Development as Freedom*. Oxford: Oxford University Press

Sinclair, A.R.E. and Fryxell, J.M. 1985. "The Sahel of Africa: ecology of a disaster," *Canadian Journal of Zoology*, 63, 987–994.

Slinger, J.H. and Vreugdenhil, H.S.I. 2007. "Can science influence policy? The inclusion of estuaries in the South African water law." In Pahl-Wostl, C., Kabat, P. and Wöltgen, J. (eds) *CAIWA 2007 Proceedings*. Basel: CAIWA, pp. 1–11.

Smutylo, T. 2001. *Crouching Impact, Hidden Attribution: overcoming threats to learning in development programs*. Draft Learning Methodology Paper, Block Island Workshop on Across Portfolio Learning, 22–24 May 2001. Ottawa, Canada: International Development Research Centre.

Smutylo, T. 2005. *Outcome Mapping: a method for tracking behavioural changes in development programs*. Institutional Learning and Change Brief No. 7. www.cgiar-ilac.org/files/publications/briefs/ILAC_Brief07_mapping.pdf.

Sommerhalter T. 2008. "Multi-stakeholder forums for co-management of pastoralism in Niger," *Nomadic Peoples*, 12, 165–177.

Start, D. and Hovland, I. 2004. *Tools for Policy Impact: a handbook for researchers*. Overseas Development Institute. www.odi.org.uk/resources/download/156.pdf.

Stringer, L.C. and Reed, M.S. 2007. "Land degradation assessment in Southern Africa: integrating local and scientific knowledge bases," *Land Degradation and Development*, 18: 99–116.

Sutton, R. 1999. *The Policy Process: an overview*. ODI Working Paper 118. London: Overseas Development Institute.

Svancara, L.K., Brannon, R., Scott, J.M., Groves, C.R., Noss, R.F. and Pressey, R.L. 2005. "Policy-driven versus evidence-based conservation: a review of political targets and biological needs," *Bioscience*, 55: 989–995.

Swanson, D. and Bhadwal, S. 2009. *Creating Adaptive Policies: a guide for policy-making in an uncertain world*. New Delhi: Sage.

Tatalovich, R. and Daynes, B.W. 1998. *Moral Controversies in American Politics: cases in social regulatory policy*, 4th edn. New York: Armonk.

TEEB. 2010. *The Economics of Ecosystems and Biodiversity: mainstreaming the economics of nature: a synthesis of the approach, conclusions and recommendations of TEEB*. UNEP and IUCN. www.teebweb.org/TEEBSynthesisReport/tabid/29410/Default.aspx.

The Living Knowledge Project. 2008. "Incorporating indigenous knowledge into your teaching." http://livingknowledge.anu.edu.au/html/educators/02_questions.htm (retrieved 1 February, 2011).

Thompson, B. and Kinne, S. 1990. "Social change theory: applications to community health." In Bracht, N. (ed.) *Health Promotion at the Community Level*. Newbury Park, CA: Sage.

Thrupp, L., Cabarie, B. and Zazueta, A. 1994. "Participatory methods in planning and political processes: linking the grassroots and policies for sustainable development," *Agriculture and Human Values*, Winter, 1–8.

Torjman, S. 2005. *What is Policy?* Caledon Institute of Social Policy. www.caledoninst.org.

Toulmin, C. and Quan, J. (eds) 2000. *Evolving Land Rights, Policy and Tenure in Africa*. London: IIED.

Turner, R.K., Paavola, J., Cooper, P., Farber, S., Jessamy, V. and Georgiou, S. 2003. "Valuing nature: lessons learned and future research directions," *Ecological Economics*, 46, 493–510.

Turner, M. and Hulme, D. 1997. *Governance Administration and Development: making the state work*. Sterling, VA: Kumarian Press.

Tyler, S. 2006. "Multi-stakeholder deliberation." In Swanson, D. and Bhadwal, S. (eds) *Creating Adaptive Policies: a guide for policy-making in an uncertain world*. Los Angeles, CA: Sage.

Ultimate Media. 2010. *Government Amends Cultural and Traditional Leaders Bill*. www.weinformers.net/2011/01/20/government-amends-cultural-and-traditional-leaders-bill/+Traditional+and+Cultural+Leaders+Bill+2010+leaders+15+out+of+21+clauses+revised+deleted&cd=1&hl=sw&ct=clnk&gl=ke.

UNDP and UNEP. 2007. *Guidance Note on Mainstreaming Environment into National Development Planning*. Nairobi, Kenya: United Nations-Gigiri.

UNDP. 1994. *Human Development Report 1994*. New York: United Nations.

UNDP. 2006. *Decentralized Governance of Natural Resources. Part 1: Manual and guidelines for practitioners*. UNDP Drylands Development Centre. New York: UNDP.

UNDP. 2008 (Emerton, L.). *Making the Economic Case: a primer on the economic arguments for mainstreaming poverty-environment linkages into development planning*. UNDP-UNEP Poverty Environment Initiative.

UNDP-UNEP. 2009. *Mainstreaming Poverty-Environment Linkages into Development Planning: a handbook for practitioners*. Nairobi, Kenya: United Nations-Gigiri.

UNRISD. 2004. *Technocratic Policy-making and Democratic Accountability*. UNRISD/RPB3e/04/3.

UNRISD. 2004. *Technocratic Policy Making and Democratic Accountability*. UNRISD conference news. UNRISD. Geneva.

van den Hove, S. 2007. "A rationale for science–policy interfaces," *Futures*, 39, 807–826.

Veit, P. (ed.) 1998. *Africa's Valuable Assets*. Washington, DC: WRI.

Waller, P., Morris, R.M. and Simpson, D. 2008. *Understanding the Formulation and Development of Government Policy in the Context of FOI*. Prepared for the Information Commisioner's Office by the Consultation Unit. London: London's Global University.

Walt, G. 1994. "How far does research influence policy?," *European Journal of Public Health*, 4: 233–235.

Wardekker, J.A., van der Sluijs, J.P., Janssen, P.H.M., Kloprogge, P. and Petersen, A.C. 2008. "Uncertainty communication in environmental assessments: views from the Dutch science–policy interface," *Environmental Science and Policy*, 11: 627–641.

Weiss, C. 1977. "Research for policy's sake: the enlightenment function of social research," *Policy Analysis*, 3 (4): 531–545.

Weiss, C.H. 2009. "Foreword." In Carden F., *Knowledge to Policy: making the most of development research*. International Development Research Centre, Ottawa, and Sage, New Delhi.

Wells, S., Makoloweka, S. and Samoilys, M. 2007. *Putting Adaptive Management into Practice: collaborative coastal management in Tanga, Northern Tanzania*. Nairobi, Kenya: IUCN. http://cmsdata.iucn.org/downloads/wells_et_al_2007_tanga_lessons_learnt.pdf.

Wenger, E. 1998. *Communities of Practice: learning, meaning, and identity*. New York: Cambridge University Press.

Wenger, E. and Snyder, W. 2000. "Communities of practice? The organizational frontier," *Harvard Business Review*, 1 January 2000.

Wiseman, J. 2010. *Dancing with Strangers: understanding the parallel universes of academic researchers and public sector policy-makers*. ANSZOG Occasional Paper No. 11. Melbourne, Australia: ANSZOG and State Services Authority. www.anzsog.edu.au/userfiles/files/News%20and%20events/SSA%20events/11_Wiseman_Final.pdf.

World Bank. 1998. World development report 1998. *Knowledge and Information for Development*. Oxford: Oxford University Press.

World Bank. 2001. *Decentralization and Governance: does decentralization improve public services delivery*? PREM Notes 55 (June). Washington, DC: Development Economics Vice Presidency and Poverty Reduction and Economic Management Network.

World Bank. 2002. *Empowerment and Poverty Reduction: a source book for poverty reduction and environmental management (PREM)*, Washington, DC: World Bank.

World Commission on Environment and Development. 1987. *Report of the World Commission on Environment and Development: our common future*. www.un-documents.net/wced-ocf.htm.

World Resources Institute. 2003. *World Resources 2002–2004: decisions for the earth: balance, voice and power*. Washington, DC: World Resources Institute. http://pdf.wri.org/wr2002_fullreport.pdf.

Worm, B., Hilborn, R., Baum, J.K., Branch, T.A., Collie, J.S., Costello, C., Fogarty, M.J., Fulton, E.A., Hutchings, J.A., Jennings, S., Jensen, O.P., Lotze, H.K., Mace, P.M., McClanahan, T.R., Minto, C., Palumbi, S.R., Parma, A.M., Ricard, D., Rosenberg, A.A., Watson, R. and Zeller, D. 2009. "Rebuilding global fisheries," *Science*, 325: 578–585.

WWF. 2009. *Indian Ocean Tuna Commission: a failure – again*. www.panda.org/wwf_news/?161442/Indian-Ocean-tuna-commission-a-failure-again. Retrieved 1 February 2011.

Wyckoff-Baird, B., Kaus, A., Christen, C., and Keck, M. 2000. *Shifting power: decentralization and biodiversity conservation*. Washington, DC: Biodiversity Support Program.

Young, E. and Quinn, L. 2002. *Writing Effective Public Policy Papers: a guide to policy advisers in Central and Eastern Europe*. Budapest, Hungary: LGI.

Young, J. 2005. *Bridging Research and Policy: the RAPID approach*. Dakar: African Economic Research Institutions and Policy Development: Opportunities and Challenges. January 28–29, 2005.

Zazueta, A. 1995. *Policy Hits the Ground: participation and equity in environmental policy-making*. USA and China: World Resources Institute.

Zeidler, J. 2010. "Some thoughts on policy and desertification: what key contributions can scientists make to policy processes in our southern African context?," *African Journal of Range and Forage Science*.

Zermoglio, F., Van Jaarsveld, A., Reid, W., Romm, J., Biggs, O., Yue, X. and Vincente, L. 2005. "The multi-scale approach." In *Millennium Ecosystem Assessment, Ecosystems and Human Wellbeing: multi-scale assessment*, vol. 4. Washington, DC: Island Press, pp. 61–84.

Zhao, B., Kreuter, U., Li, B., Ma, Z., Chen, J. and Nakagoshi, N. 2004. "An ecosystem service value assessment of land-use change on Chongming Island, China," *Land Use Policy*, 21: 139–148.

Zirimwabagabo, I. 2006 "Gender-based violence bill passes in Rwandan Parliament." *UNIFEM News*, 7 August. www.unifem.org/news_events/story_detail.php?StoryID=502.

Zukosky, M.L. 2007. "Making pastoralist settlement visible in China," *Nomadic Peoples*, 11: 107–133.

INDEX

For Product Safety Concerns and Information please contact our EU representative GPSR@taylorandfrancis.com Taylor & Francis Verlag GmbH, Kaufingerstraße 24, 80331 München, Germany